Advances in

Insect Physiology

Volume 20

Advances in
Insect Physiology

edited by

P. D. EVANS
and V. B. WIGGLESWORTH

*Department of Zoology, The University
Cambridge, England*

Volume 20

1988

ACADEMIC PRESS

Harcourt Brace Jovanovich, Publishers

London   San Diego
New York   Boston
Sydney   Toronto   Tokyo

ACADEMIC PRESS LIMITED.
24/28 Oval Road
London NW1 7DX

United States Edition published by
ACADEMIC PRESS INC.
Orlando, Florida 32887

British Library Cataloguing in Publication Data

Advances in insect physiology.—Vol. 20
    1. Insects—Physiology—Periodicals
    595.7'01'05   QL495
ISBN 0-12-024220-6
ISSN 0065-2806

Typeset by Eta Services Ltd, Beccles, Suffolk
and printed in Great Britain by St Edmundsbury Press, Bury St Edmunds, Suffolk

Contributors

A. D. Blest

Developmental Neurobiology Groups, Research School of Biological Sciences, Australian National University, PO Box 475, Canberra City, ACT 2601, Australia

J. N. Carr

Department of Anatomy and Neurobiology, Washington University School of Medicine, 660 S Euclid Avenue, St Louis, MO 63110, USA

T. M. Casey

Department of Entomology and Economic Zoology, New Jersey Agricultural Experiment Station, Cook College, Rutgers University, New Brunswick, NJ 08903, USA

J. L. Gould

Department of Biology, Princeton University, Princeton, NJ 08544, USA

D. B. Sattelle

ARFC Unit of Insect Physiology and Pharmacology, Department of Zoology, University of Cambridge, Downing Street, Cambridge CB2 3ET, UK

P. H. Taghert

Department of Anatomy and Neurobiology, Washington University School of Medicine, 660 S Euclid Avenue, St Louis, MO 63110, USA

W. F. Towne

Department of Biology, Kutztown University, Kutztown, PA 19530, USA

J. B. Wall

*Department of Anatomy and Neurobiology, Washington University School of
Medicine, 660 S Euclid Avenue, St Louis, MO 63110, USA*

D. Yamamoto

*Neuroscience Division, Mitsubishi-Kasei Institute of Life Sciences,
11 Minamiooya, Machida, Tokyo 194, Japan*

Contents

The Turnover of Phototransductive Membrane in Compound Eyes and Ocelli

A. David Blest

Developmental Neurobiology Groups, Research School of Biological Sciences, Australian National University, PO Box 475, Canberra City, A.C.T. 2601, Australia

ADVANCES IN INSECT PHYSIOLOGY VOL. 20
ISBN 0–12–024220–6

1 Introduction

Almost simultaneously in the late 1960s, Eguchi and Waterman (1967) and White (1967a, b) revealed that the phototransductive membranes of the eyes of arthropods are in a state of flux, and exhibit accelerated turnover in response to light, an issue that was explored quantitatively by White and Lord (1975). In a seminal pioneer study, Young (1967) discovered that packages of discs are shed daily from the tips of the rod outer segments of vertebrates. Later, Blest (1978), Nässel and Waterman (1979) and Williams (1982b) showed that for a nocturnal spider, a crab and a locust, respectively, turnover events are locked to normal daily cycles of natural illuminance, and that they can proceed in such a way as to tailor the membrane architecture of a receptor for the special demands of day and night vision.

The extent to which less obviously dedicated plasma membranes are turned over was not appreciated at the time of the earlier studies. Retinal physiologists assumed that cellular specializations for the capture of photons were architecturally stable—an intuitive conclusion that recent applications of optical theory to problems of retinal design might seem to make all the more compelling (Snyder, 1979; Land, 1980). Nevertheless, the assumption of stability was wrong and its simplicity misleading.

Lability of transductive membrane expressed by complex temporal patterns of turnover raises a number of problems that can be summarized along the following lines. To trap photons and convert their energy to electrical signals, the plasma membranes of photoreceptors contain photopigments* – in the case of arthropods, rhodopsins, xanthopsins (Vogt, 1983, 1984; Vogt and Kirschfeld, 1984; Kirschfeld, 1986), and so forth – as their major integral proteins. The photopigments of invertebrate photoreceptors, and their transitions consequent upon illumination are reviewed by Hamdorf (1979) and Stavenga and Schwemer (1984). Because photon capture is statistically inefficient, phototransductive membrane must be both amplified and compacted. Typically, an arthropod photoreceptor includes a specialized

---

*Invertebrate photopigments will be discussed generically as "rhodopsins", and their photoproducts as "metarhodopsins", for reasons of economy. In contexts in which a distinction is irrelevant, mixtures of "rhodopsin" and "metarhodopsin" will be described as "rhodopsin".

region where the plasma membrane is organized as densely-packed microvilli interposed in a light-path as a coherent domain. Optical constraints on the shapes of such domains and the physical properties of their boundaries can be rigorous (Snyder, 1979). Radical turnover phenomena impose formidable demands upon a photoreceptor: what decides the overall polarity of such a cell, and where upon its surface will the transductive domain be situated? Given that a transductive domain is in a continuous state of flux, what ensures that it will maintain a functionally critical shape throughout numerous cycles of turnover? Why is turnover necessary in the first place, and what would go wrong if it did not occur? How are the events of turnover regulated physiologically?

Each of these problems must eventually be analysed in molecular terms, and one role of this review will be to attempt some conceptual refinement of the field so that the right questions can be asked. The approach will be eclectic and comparative; turnover strategies vary greatly between different arthropods. Perhaps the evolutionary considerations that should never be far from our minds justify acceptance of *Limulus*, spiders and crustaceans as honorary insects for our present purposes. The large literature on the turnover of the phototransductive membranes of vertebrates can only be mentioned briefly here in a few appropriate contexts.

2 Organization of arthropod photoreceptors

2.1 REGIONAL DIFFERENTIATION OF FUNCTION

The photoreceptors of arthropods are complex cells that serve several different functions simultaneously. Not only must they trap photons and convert their energy to electrical signals, but the resultant information must be conveyed to the next processing stages of the visual system, and transmitted. An arthropod photoreceptor may comprise:

(1) A somatic region or *soma* that contains the nucleus;

(2) A photon-trapping region in which the plasmalemma is disposed as slender, densely-packed microvilli—the *receptive segment*;

(3) *An intermediate segment* which, like the soma, contains organelles such as endoplasmic reticulum, Golgi complexes, mitochondria, etc.;

(4) A slender *axon* that conducts signals to the visual neuropil;

(5) A *synaptic terminal* that transmits them.

These separate regions are not always clearly defined and can be arranged in a variety of ways (Fig. 1). In most instances, however, a photoreceptor will

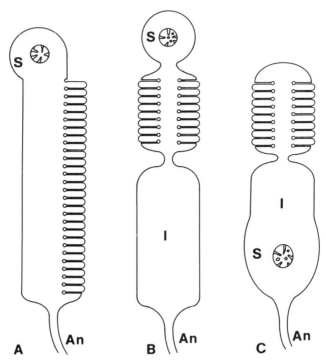

Fig. 1   Three diagrams to show typical arrangements of phototransductive microvilli
on the cell surfaces of single arthropod photoreceptors. (A) A typical insect. Microvilli occupy a restricted region of the cell. The rhabdomere to which they contribute is
as long as 300 μm in some cases, e.g. locusts. There is no discrete intermediate segment, and the soma is distal. (B) A receptor of a secondary eye in a nocturnal spider.
The soma is distal, and the intermediate segment lies proximal to the receptive segment. (C) A receptor of the principal eye in a nocturnal spider. The intermediate segment lies proximal to the receptive segment, but the "soma" (represented by the
region of the nucleus) lies more proximally still. In both (B) and (C) the microvilli of
the receptive segment may cover all faces of the plasmalemma. S, soma; I, intermediate segment; An, axon.

possess an unequivocal polarity and a strictly-determined architecture. The
two regions with which this review is mainly concerned—the receptive and
intermediate segments—are especially complex because of their multiple
functions. To understand why, it is necessary to remember that a photoreceptor is merely an exceptionally elaborate sensory neurone. If, for
example, the receptive and intermediate segments are interposed between the
soma and the axon (Fig. 1b), fast and slow axonal transport must convey
materials centrifugally towards the synaptic terminal, carrying them through
the two most active regions of the cell, for the receptive segment contains a

local enzymatic machinery for transduction and adaptation, while the intermediate segment must be able to sustain turnover of the transductive membrane. Components of the latter are synthesized by an extensive endoplasmic reticular system and, since most of these cells internalize membrane for degradation, they contain an extensive lysosomal apparatus. Traffic of membrane to and from the microvilli, however, demands that there also be a presumably independent system for radial transport.

Phototransductive microvilli are said to be organized as a *rhabdomere*, a domain contributed by a single photoreceptor, or as a *rhabdom*, an optically discrete domain composed of the rhabdomeres of several contiguous cells. Figure 2 represents the rhabdom of *Leptograpsus*, a crab which has been used extensively for studies of turnover.

2.2.   THE COMPOSITION OF PHOTORECEPTOR MICROVILLI

The phenomenon of turnover requires us to consider what components of a microvillus are destroyed and replaced.

2.2.1   *Integral membrane proteins*

Rhodopsins are the major components of a microvillus, and have been shown to provide the bulk of the P-face particles in freeze-fracture preparations (Boschek and Hamdorf, 1976; Schinz *et al.*, 1982). De Couet (1984), studying crayfish photoreceptors, has demonstrated the presence of a minor population of glycoproteins of higher molecular weight with complex oligosaccharide chains. Their functions are unknown, although it is reasonable to suppose that some may anchor components of the microvillar cytoskeleton. The sites of ion channels in microvillar membranes are a matter of dispute, and estimates of their densities rely upon indirect evidence. For instance, it can be argued on statistical grounds that there need be no more than one sodium channel per microvillus (Hamdorf and Kirschfeld, 1980). Whatever the roles of proteins other than rhodopsin in the microvillar membranes, it is worth noting that some (or perhaps one) of them can be visualized as rows of large P-face particles extending longitudinally in linear arrays along each microvillus (Boschek and Hamdorf, 1976; Blest *et al.*, 1982a).

2.2.2   *Cytoskeletal and associated proteins*

Microvillar architecture is best understood for the avian intestinal brush-border (Mooseker, 1983). Bundles of actin microfilaments are cross-linked,

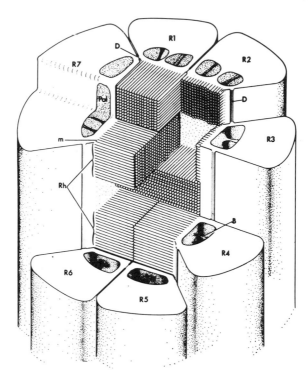

Fig. 2   Schematic diagram showing the organization of the fused rhabdom of a crab, *Leptograpsus*. The fused rhabdom, Rh, is composed of microvilli (m) contributed by seven retinula cells, R_{1-7}, that are linked by desmosomes, D. The complex inter-digitation of the blocks of rhabdomeral microvilli contributed by the retinula cells is typical of many crustaceans. In insects with fused rhabdoms the organization, with few exceptions, is simpler. Saccules derived from the smooth endoplasmic reticulum form a "palisade" (Pal) around the rhabdom; the refractive index differences between the two components ensure that the rhabdom can act as a light guide. Exchanges of membrane between rhabdom can act as a light guide. Exchanges of membrane between rhabdom and soma involve transit of materials across the narrow cyto-plasmic bridges (B) that span the palisade. (From Stowe, 1980.)

and are also bound to the plasma membrane by cross-bridges. There are associated, regulatory proteins, and the Ca^{2+}-regulating protein, calmodu-lin. Arthropod photoreceptor microvilli also contain a cytoskeleton. Varela and Porter (1969) first illustrated a single axial filament lying in the lumen of a microvillus of the photoreceptor of a bee.

Blest *et al.* (1982a, b) stabilized similar single filaments linked to the plasma membrane by side-arms in the blowfly *Lucilia*. Either Ca^{2+} was chelated from both a pretreatment medium and a primary fixative solution, or receptors were pretreated with specific inhibitors of cysteine proteases. In both cases, Mg^{2+} ions were supplied to stabilize side-arms. A subsequent study on crayfish photoreceptors implied that a single axial filament must consist of F-actin (de Couet et al., 1984), but freeze-substitution data from crayfish suggest that a microvillus may contain several microfilaments (J. Usukura, unpublished data). The wide (80–140 nm), interdigitated microvilli of some nocturnal spiders reveal microfilaments closely associated with the plasma membranes (Blest and Sigmund, 1985). It may be noted that if the slender (33–65 nm) microvilli of Diptera and crustaceans were to contain similarly disposed microfilaments, it is unlikely that conventional techniques of transmission electron microscopy could resolve them, because of the unsatisfactory relationship between section thickness and microvillar diameter (Blest and Sigmund, 1985).

The presence, extent and roles of an actin cytoskeleton will be seen to be important when we come to discuss models for the assembly of microvilli during turnover, and for mechanisms that selectively remove integral membrane proteins from the microvilli.

An F-actin cytoskeleton implies the presence of other cytoskeletal proteins, following the precedent set by the avian enterocyte, but their contribution has not been established. Calmodulin has been shown to be largely localized to the rhabdomeral microvilli of squid, *Drosophila*, the blowfly *Lucilia*, crab and crayfish (de Couet et al., 1986). Although it may merely buffer the microvilli against the high Ca^{2+} fluxes that occur during phototransduction (Kirschfeld and Vogt, 1980), it is more likely that it enacts multiple roles in contexts that will become apparent below.

2.2.3 Transductive systems

In arthropods, there is little direct evidence concerning specific enzymes of the transduction cascade, or where they may be localized. One identified enzyme can be cited as an informative paradigm. Trowell (1984 and in press) has characterized a phosphoprotein phosphatase in squid and crab photoreceptors, distinct from but resembling the calcineurin of vertebrate neural tissue (Pallen and Wang, 1985). Ultrastructural cytochemistry shows it to be localized in the microvilli and presumptively rhodopsin-containing membranes elsewhere. Yet it is not an integral membrane protein. It is calmodulin-dependent and is probably bound to the plasmalemmal undercoats (Blest and Eddey, 1984) of the microvilli and the regions of plasmalemma flanking them. It serves as an example of proteins whose fate during turnover must be exactly regulated, but whose exchanges cannot be inferred from ultrastructural examination of membrane traffic alone.

There is probably substantial homology between invertebrates and vertebrates for many proteins of the transduction cascade; Tsuda et al. (1986), for example, showed homology of the GTP-binding protein (G-protein, transducin) over a wide range of animals.

2.2.4 *Proteases*

Preservation of the cytoskeleton of microvilli by specific inhibitors of cysteine proteases implies their presence within both the transductive apparatus and the photoreceptor cytosol (Blest *et al.*, 1982a, b, 1984a; Blest and Eddey, 1984; Blest and Sigmund, 1985; de Couet *et al.*, 1984). Significantly, 80% of the high complement of similar enzymes (calpains) in the retinae of a vertebrate is contained in the photoreceptor outer segments (Tsung and Lombardini, 1985). In neither case do we yet know what these enzymes are doing, but it is reasonable to hold them in reserve as possibly concerned in turnover events.

To categorize rhabdomeral components in such a manner may seem over-elaborate, but it is intended to suggest that there is much to evaluate beyond the simple ultrastructural phenomenology that for the most part is all that has been attempted, and with which subsequent sections must largely be concerned.

3 Phenomenology of turnover: adjustments to the volumes of rhabdoms

Historically, light- or darkness-induced changes to the dimensions of rhabdomeres or rhabdoms were studied before the nature of the daily cycles of the turnover of transductive membrane was appreciated. From a number of early studies, one may mention those of Sato *et al.* (1957; mosquito), Tuurala and Lehtinen (1971a, b, 1974; the terrestrial crustacean *Oniscus*); Behrens (1974) and Behrens and Krebs (1976; *Limulus*), Brammer and Clarin (1976) and White and Lord (1975; mosquito), and of Meyer-Rochow and Waldvogel (1979; the mycetophilid fly *Arachnocampa*). They anticipated a subsequent generation of studies, and were seminal in provoking them. In terms of the background that they provided, the effects of light and darkness can best be discussed in the following way.

3.1 QUANTITATIVE EFFECTS OF PROLONGED ILLUMINATION

White and Lord (1975) examined larval rhabdoms of the mosquito *Aedes*. Fourth instar larvae were exposed to defined levels and durations of illumination, during which the sizes of rhabdoms diminished. During

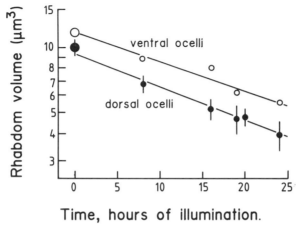

Time, hours of illumination.

Fig. 3   Diminution of volume in larvae of white-eyed mutants of the mosquito, *Aedes aegypti*, reared in darkness and then exposed to continuous illumination of $1 \, mW/cm^2$. Rhabdom volume decreased exponentially over some 24 hours, after which it stabilized. (From White and Lord, 1975.)

prolonged dark adaptation, their sizes increased. Rates of diminution when illuminated related to light intensities (Fig. 3).

The rhabdoms of larvae exposed to various light intensities for 72 hours arrived at steady state dimensions proportional to the levels of illumination (Fig. 4). Nevertheless, such rhabdoms seemed to be in a state of flux; endocytosis of their microvillar membranes was seen to be continuing, and must, therefore, have been balanced by membrane synthesis. White and Lord (1975) concluded that changes of rhabdom volume under their experimental conditions reflected an influence of light upon phototransductive membrane turnover, and that a particular steady state represented an equilibrium between the processes of membrane catabolism and those responsible for renewal. Their findings characterized the variable balance between membrane breakdown and synthesis, and provided the only meticulous quantitative analysis of the influence of light upon the turnover of phototransductive membranes that we have.

3.2   QUANTITATIVE EFFECTS OF DAILY CYCLES OF ILLUMINATION

As the results summarized in the previous section would suggest, rhabdoms decrease in size at natural dawn, with the onset of light, and increase at dusk (Williams, 1982b, locust; Stowe, 1981, crab; Blest, 1978, the spider *Dinopis*; Nässel and Waterman, 1979, crab). More surprisingly, in locust, at least, the

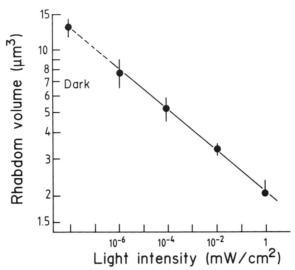

Fig. 4   The relationship of steady-state rhabdom volume to light intensity in larval white-eyed mutants of the mosquito, *Aedes aegypti*. Initially dark-adapted larvae were exposed to various light levels for 72 hours; each point represents the steady state reached after exposure. (Modified from White and Lord, 1975.)

rhabdoms of animals held in darkness over the dawn period also decrease in size, although more slowly (Williams, 1982b). The effects of different protocols of illumination on the diameters of rhabdoms in locust are summarized in Figure 5. Differences between the cross-sectional areas (and, therefore, volumes) of rhabdoms between the day and night states can be considerable (Fig. 6).

Blest (1980) and Blest *et al.* (1984b) have suggested that dramatic daily changes in the dimensions of rhabdoms are associated with arthropods that are active over a wide range of illuminances. Strictly speaking, this conclusion is only indicated for compound eyes, for the ocellar retinae of numerous species of purely nocturnal spider drastically reduce the amounts of rhabdomeral membrane in their receptors during the day, and may even eliminate it altogether (Blest, 1985).

It is probable, however, that most insects make little adjustment to the sizes of their rhabdomeres; Williams (1982b) found no significant changes in the CSAs of rhabdomeres R_3 or R_7 in the course of a 12:12 hour L/D cycle in *Lucilia*. Similarly, Blest and Maples (1979) were able to observe directly that little membrane is shed in the course of a daily cycle from rhabdomeres of the salticid spider *Plexippus*.

Blest (1980) suggested that it is useful to distinguish between "variable

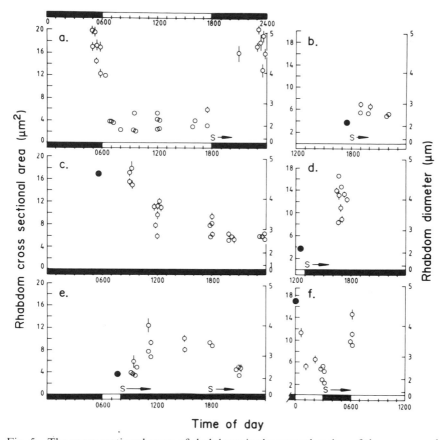

Fig. 5 The cross-sectional areas of rhabdoms in the central region of the compound eyes of *Locusta*, in relation to times of day and to schedules of light and darkness. A point represents the mean size of 15–30 rhabdoms from one insect; vertical bars extend two standard errors of the mean. Large filled circles indicate mean sizes on a normal cycle before the onset of an experimental cycle of illumination. S→ indicates points in time at which microvillar assembly was seen, ultrastructurally, to take place. Locusts were exposed to a 12:12 hour light/dark cycle for at least 14 days before an experiment. (*a*) Normal light/dark cycle. (*b*) Light continued into the "dark" phase of the cycle. (*c*) Darkness maintained over the "light" phase of a normal cycle. (*d*) Darkness imposed 5 hours before its onset in a normal cycle. (*e*) Darkness imposed 10 hours before its onset in a normal cycle. (*f*) Illumination imposed for 3 hours after midnight. (Modified from Williams, 1982a.)

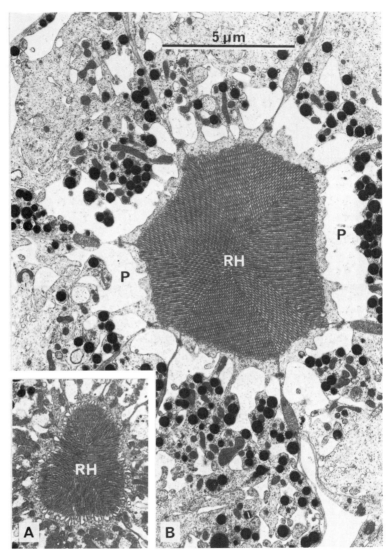

Fig. 6 Transverse sections of the fused rhabdoms of an unidentified Panamanian tettigoniid katydid, taken just proximal to the termination of the crystalline cone, at mid-day (A) and at midnight (B), on a normal environmental light cycle. In addition to the enlargement and reorganization of the rhabdom, (RH), at night, reconstruction of the saccular palisade that surrounds it is also evident. The katydid is a nocturnal, predatory or scavenging carnivore that hides in rolled-up leaves during the day. It nevertheless exhibits marked visual responses when disturbed from its refuge. Scale on (B) (Both micrographs × 8000.)

size" and "stable" rhabdoms, but it is clear that cases such as that of the tipulid fly, *Ptilogyna*, described by Williams (1980b) and Williams and Blest (1980) lie somewhere in between. In all cases, there is ultrastructural evidence that turnover is probably proceeding to some extent at all times of day, possibly vigorously. R_{1-6} in *Drosophila* are "stable", but the large amounts of mRNA for the R_{1-6} rhodopsin implied by Zuker *et al.* (1985) suggest that the rhodopsin–membrane complex may turn over rapidly. It is not known how rapidly endocytosis proceeds; the data of Hubbard and Stukenbrok (1979) from hepatocytes (discussed in more detail by Blest and Eddey, 1984) also imply that it is fast, so that endocytotic profiles observed by electron microscopy in photoreceptors are assumed to "freeze" a rapid process.

3.3. EVOLUTIONARY PATTERNS OF ADJUSTMENT OF RHABDOM VOLUME

Although the changes in rhabdomeral dimensions in the course of a daily cycle can be spectacular, there is no reason to suppose that transductive membrane exchanges are any greater in relative area than those of plasma membranes in general, and they may even be less (Nässel and Waterman, 1979). The latter authors noted the high rates of pinocytosis of plasma membrane by *Amoeba*. Roughly 20%/hour has been measured for rat adipocytes in culture (Gibbs and Lienhard, 1984), and this value may be conservative for other cells.

It seems reasonable to conclude that the turnover of phototransductive membrane does not represent an evolutionary novelty *per se*, but that it has been tailored by natural selection to match specific ecological requirements, and the consequent optical demands.

It is worth noting that the daily shedding of phagosomes from the tips of the outer segments of the photoreceptors of vertebrates (Young, 1976) must, at least in rods, represent a rather low relative rate of membrane turnover, because only a small region of the outer segment is lost, and the bulk of an outer segment is stable over periods of days.

4 The control of phototransductive membrane turnover

4.1 LOCAL CONTROL BY STATES OF ILLUMINATION

The compound retinae of insects do not receive an efferent innervation, although it should be remembered that there are many efferent fibres able to supply photoreceptor terminals in the first optic neuropil, the lamina

(Strausfeld, 1976). Events of turnover could, therefore, be determined by an endogenous, central control, exerting an influence remote from the retina itself.

In insects, the situation has only been studied intensively in locusts (Williams, 1982b, 1983). A masked region of a compound eye fails to shed membrane normally at dawn, and uniquely assembles new membrane at night if the unmasked portion of the eye remains illuminated. These results demonstrate that the events immediately triggered by states of illumination are controlled at least at the level of the individual ommatidium, and probably at that of the single photoreceptor. Nevertheless, endogenous factors are also involved. Locusts held in continuous darkness for 12 hours from "dawn" (so that they do not experience normal "day") exhibit a slow diminution of rhabdom diameter (Fig. 5). Animals allowed 2 hours of illumination at "dawn" and then returned to darkness lose rhabdomeral membrane in the normal way, but in premature darkness fail to re-assemble a "night" rhabdom to more than circa 50% of its normal diameter. In continuous darkness, such a premature "night" rhabdom shows ultrastructural evidence of renewed synthesis at the time at which dusk would have occurred, but fails to reconstitute a structure of the predicted, nocturnal size.

This complex situation allows the possibility of control either by efferents acting at the level of the lamina, or by endocrine pathways. In the locust, the evidence does not definitely demand either mechanism; cyclical phenomena could be an outcome of rhythms intrinsic to the receptors themselves, amplified by the complex supply-and-demand effects that might be generated by the cyclicity of environmental signals.

Illumination and darkness regulate turnover in the photoreceptors of the crab *Leptograpsus* in essentially the same way (Stowe, 1981). The retinae of *Leptograpsus* can be maintained in an oxygenated saline solution for several hours (Stowe, 1982). Induction of synthesis and assembly by early darkness proceeds normally in cultured retinae, and is clearly not dependent upon central inputs, either neural or hormonal. Isolated retinae with laminae attached undergo normal breakdown of "night" rhabdoms in response to light, but slow diminution of rhabdoms in continuous darkness fails to occur. The latter process, therefore, is probably under central control.

4.2   IONIC BALANCE

Stowe (1983a) investigated the putative role of calcium in the process of rhabdom reduction in *Leptograpsus*, using retinae *in vitro*. She failed to demonstrate any effect of external Ca^{2+} levels as high as 10 mM, even in the presence of 20 μm A23187, a Ca^{2+} ionophore. Massive breakdown was precipitated by as little as $1\,\mu M$ monensin, a sodium/potassium ionophore. High potassium concentrations in the external medium reduced the diameters of "night" rhabdoms linearly with respect to log K_{ext}^{+}. From the Nernst equation, a potassium-induced depolarization of 10 mV would correspond to a reduction of rhabdom diameter of $0\cdot3\,\mu m$, and a depolarization of the photoreceptor resting potential of around 70 mV would be required to convert a "night" to a "day" rhabdom. Although these experiments show that depolarization is sufficient to induce breakdown, they do not prove that it is necessary *in vivo*. Martin and Hafner (1986) showed, from meticulous counts of lysosomes in crayfish, that Ca^{2+} influences rates of degradation. It seems to be implied that control of the internalization of microvillar membrane may be independent of Ca^{2+} fluxes, but that the degradation of internalized membranes may be partially controlled by them. The issues involved are clearly complex, and it is difficult at present to envisage a satisfactory experimental approach to them.

4.3   EFFERENT CONTROL OF MEMBRANE TURNOVER IN *LIMULUS*

Barlow *et al.* (1977, 1980) described sensitivity changes of the photoreceptors of *Limulus* that followed a circadian rhythm, and which were regulated by efferent inputs that end as synaptoid terminals in the retinae (Barlow, 1983). The changes observed related both to the photomechanical adjustments previously described by a number of authors (Behrens, 1974; Behrens and Krebs, 1976; Miller, 1975; Miller and Cawthon, 1974), and to the consequences of phototransductive membrane turnover (Barlow and Chamberlain, 1980).

Subsequently, efferent retinal terminals were shown to synthesize and release octopamine (Battelle *et al.*, 1982), the transmitter determining the course of phototransductive membrane turnover. Efferent control of this kind is unlikely to be implicated in the regulation of membrane turnover in the photoreceptors of higher arthropods, although the possibility cannot be excluded. At present, it should be noted that, despite an enormous volume of published work on the ultrastructure of the retinae of the compound eyes of both insects and crustaceans, no account describes synaptoid endings within the retinae themselves of the kind found in *Limulus*.

The distribution of synaptoid endings within the retinae of spiders (Blest, 1985) is suggestive. They are found in representatives of several nocturnal families with low acuity vision and radical turnover strategies. They are certainly absent in the diurnal Salticidae (which possess high acuity vision in both the accessory and principal eyes), and in the Dinopidae which are highly specialized for nocturnal vision (Blest and Land, 1977; Blest *et al.*, 1980b). Furthermore, in locust, Williams (1982c) demonstrated a precise local control of turnover that is not consistent with regulation by an anatomically diffuse efferent input. The implication is that efferent control was replaced by local mechanisms as the various eyes moved during evolution to sophisticated levels of image processing. Nevertheless, it is still not impossible that in the compound eyes of insects the performance of individual receptors is modulated by efferent inputs to their terminals in the first optic neuropil, the lamina, where there is known to be an efferent system of unknown function but of some complexity (Strausfeld, 1976). Some events in the turnover routines of the photoreceptors of *Leptograpsus* seem to demand an efferent or neurohormonal input (Section 4.2).

5   Physiological and optical consequences of turnover

Differences between the day and night dimensions of rhabdomeres or rhabdoms have optical consequences for the capabilities of the eyes in which they reside. Here, "optics" includes the photon flux that will be received by a single absorptive unit (a rhabdom or a rhabdomere), and the absorption coefficient of the amplified and condensed phototransductive membrane that will determine how efficiently the photon flux is utilized.

Rhabdoms (or rhabdomeres) whose shapes and volumes change during a daily cycle will be considered first. An uncomplicated case is offered by the posterior median eye of the nocturnal spider *Dinopis*. The eye consists of a simple ocellus that delivers images to an untiered retina with large receptive segments some 20–22 μm in diameter; transductive membrane is arranged around the boundaries of the receptive segments as "rhabdomeral networks" (Blest, 1978, 1985; Blest and Land, 1977). Analysis is simplified by the absence of screening pigment migrations, since the photoreceptors of spiders do not possess "longitudinal pupils" composed of intracellular pigment granules of the kind found in higher Diptera and many other arthropods (Kirschfeld and Franceschini, 1969). It is possible in *Dinopis* to calculate the increment in photon absorption that should result solely from the synthesis of new membrane at night on an assumption that day and night membranes are identical, and that they are equivalently packed. A day rhabdom has been estimated to capture some 6% of photons incident upon a single receptor, and a night rhabdom some 74% (Blest, 1978). Intracellular recordings by Laughlin *et al.* (1980) from receptors in the day state produced a consonant estimate, but for technical reasons it proved impossible to complement it for the night condition. Estimates of performance from anatomical and optical data alone depend upon values assumed for the absorption coefficients of rhabdoms, and they are uncertain.

The proportion of photons accepted by a rhabdom (or rhabdomere) is a function of its length, x, and equals $1 - e,^{-kx}$ k being the absorption coefficient of the rhabdom or rhabdomere (Blest and Land, 1977). Values of k may be supposed to be diverse, because they will depend *inter alia* upon the density of photopigment molecules in microvillar membranes. k has been given a value of $0.01/\mu m^{-1}$ for various arthropods (Kirschfeld, 1969; Hays and Goldsmith, 1969), but Bruno *et al.* (1977) obtained $0.0067/\mu m^{-1}$ for the rhabdoms of a lobster. Estimates of performance for rhabdoms of a single species in their day and night states derived from optical and anatomical data alone assume that both the dimensions of microvilli and the densities of photopigment within their membranes are constant, and it cannot be assumed that this will always be true.

An analysis of the consequences of turnover strategies in the compound eyes of locusts by Williams (1982b, 1983) following a brief survey by Horridge *et al.* (1981) dealt satisfactory with this problem.

(1) There was no evidence that the nature of the transductive membrane differs between day and night, despite the radical reconstruction of the rhabdom that occurs at each transition between them. Firstly, freeze-fracture preparations show that particles on P-faces have identical densities in the two states (Williams, 1982b), such populations being assumed to represent rhodopsin concentrations. Secondly, the absolute sensitivities to point sources accurately disposed on-axis with respect to ommatidia from which recordings are being made from single cells penetrated by microelectrodes are the same in both day and night states. They were measured as frequencies of photon capture, a single quantal event yielding a discrete electrical signal (a "bump"). Quantum capture efficiency does not change.

(2) The angular sensitivities of ommatidia, however, do change (Wilson, 1975). Because light received by a fused rhabdom is pooled amongst its constituent receptors, a recording from a single cell can be used to evaluate the angular sensitivity of the ommatidium to which it belongs. The fields of view of an ommatidium, $\Delta\rho$, are complicated by two factors, illustrated in Figure 7. The light accepted by the tip of a rhabdom will be constrained by the state of a field stop; also, a saccular "palisade" that sleeves the rhabdom throughout its length will affect the extent to which light focused on its tip will be transmitted down it by internal reflection, the palisade being well-developed at night and considerably reduced during the day. For the locust *Valanga*, angular acceptance at 50% sensitivity increases from $1.7°$ to $4.7°$ during the day-to-night transition and in *Locusta* from $1.9°$ to $4.9°$.

(3) In both *Valanga* and *Locusta*, sensitivity to an extended source increases by at least one log unit during the first 1–3 hours after "dusk"; 0.6 log

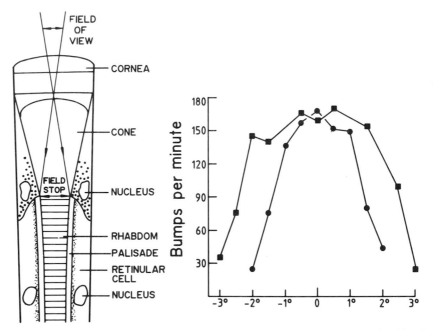

Fig. 7 (A) The optics of a locust ommatidium. The field of view is determined by the cross-sectional area of the distal tip of the fused rhabdom, and by the state of a field-stop provided by pigment granules of primary pigment cells that surround the distal end of the rhabdom. The organization of the palisade (which changes between night and day) modulates the light-guide properties of the rhabdom. (B) The angular sensitivity function measured by intracellular recording from a single photoreceptor of *Locusta* at "dusk" (●), and at five hours after "dusk" (■). The stimulus was a point source of monochromatic light (wavelength 413 nm) subtending 0·1° at the eye. Responses were measured as frequencies of photon capture; a quantal event is indicated by a "bump" in the electrophysiological record. (From Williams, 1983.)

units of this increase (amounting to a × 3·9 increment) can be attributed to the combined effects of changes to the field stop, and enlargement of the rhabdom.

The outcome of this analysis is disappointing for cell biologists. There is nothing to suggest that the properties of the transductive membrane differ between night and day. The conclusion that ommatidia at night have higher sensitivities but lower resolution than they do during the day is nevertheless interesting functionally, because it parallels the well-known higher-order pooling of rod responses after dark-adaptation in the retinae of vertebrates (Pirenne, 1967).

A similar, but less detailed analysis of the consequences of changes of rhabdom volume, compounded with the effects of screening pigment migrations has been conducted on *Limulus* (Barlow *et al.*, 1980). The results seem essentially compatible with those of Williams (1983).

6 Biogenesis of transductive membrane

6.1 BIOGENESIS

Whittle (1976) drew attention to the elaborate endoplasmic reticular (ER) specializations found in invertebrate photoreceptors; Itaya (1976) also noted continuities between saccules of smooth endoplasmic reticulum (SER) and the bases of microvilli in a shrimp. Both authors considered that these massive amounts of SER might relate to the renewal of rhabdoms and the formation of new microvilli.

Before discussing the ambiguities inherent in our current understanding of the biogenesis of arthropod microvillar membrane, it is helpful to summarize the pathway that has been carefully determined for the renewal of outer segment membrane in the rods of vertebrates (Papermaster and Schneider, 1982; Bok, 1985).

(1) Opsin is inserted at synthesis into ER membrane;

(2) Some glycosylation near to the NH_2 terminus is achieved in the RER;

(3) Trimming and further addition of sugar residues is accomplished by traffic through the Golgi;

(4) Vesicles incorporating opsin are conveyed to the base of the outer segment, where they are incorporated into nascent disc membrane via a route that is still incompletely resolved but clearly implicates the periciliary ridge complex;

(5) The chromophore, 11-cis retinaldehyde, is not added in either the RER or the Golgi;

(6) It would seem unlikely, therefore, that the presence of chromophore is a necessary stimulus for the synthesis of opsin.

In arthropods, continuity between SER and microvillar bases is so frequently observed that it has seemed natural to propose that microvilli may be assembled by membrane flow and re-organization rather than by the addition of small vesicles (Itaya, 1976; Stowe, 1980). This model seems to be implied by observations on the differentiation of RER in *Leptograpsus* during the period before dusk (Stowe, 1980): saccular sheets of RER undergo massive transformation to saccules and large tubules of SER, with which they can be seen in continuity. Such SER tubules can either be observed unmodified, in transit across the bridges of the palisade, or to undergo further transformations beforehand. In both cases, membrane whorls of characteristic appearance, termed "doublet ER" by Stowe (1980), are produced; seemingly, it is these that are transformed into microvilli by a topologically complex change in conformation whose exact sequence is not understood.

In *Leptograpsus*, the Golgi is both conspicuous and active around the time of synthesis, but the total removal of the "old" rhabdom that precedes reassembly may require an injection of lysosomal enzymes, although, if so, their source is unknown (Section 5.2.2). Bypassing of the Golgi, however, might be held either to imply that rhodopsin is not glycosylated, or that sugar residues added in the RER are not subsequently modified by trimming. There is very little hard evidence for this. De Couet (1984) failed to obtain lectin binding to the major rhodopsin of the crayfish *Cherax*, although there was a weak Schiff reaction on gels. de Couet and Tanimura (1987) obtained more convincing evidence that the rhodopsin of R_{1-6} photoreceptors in *Drosophila* is also non-glycosylated. Conversely, Hafner (1984) and Hafner and Tokarski (1986) demonstrated incorporation of ^{3}H-mannose and ^{3}H-fucose into the predominant rhodopsin of the crayfish *Procambarus*. The complex sugar chains of the higher molecular weight glycoproteins of *Cherax* rhabdoms revealed by lectin binding also imply transit through the Golgi. Although a possible return pathway from the Golgi to the ER has been suggested (Rothman, 1981), evidence that it exists has not been found so far (Brands *et al.*, 1985). On the other hand, at least one plasma membrane protein is known to reach its destination directly from the ER, without passage through the Golgi (Brands *et al.*, 1985), and Sluiman (1984) has shown from ultrastructural evidence that in a green algae, *Cylindrocapsa*, biogenesis of the plasma membrane also bypasses the Golgi compartment.

It is pertinent that rhodopsins of arthropods are heterogeneous. Monoclonal antibodies to the rhodopsin of retinula cells 1–7 in *Cherax* do not cross-react with rhodopsins of *Drosophila*, the blowfly *Lucilia*, *Leptograpsus* or squid (de Couet and Sigmund, 1985). Failure to cross-react must go further than the sequence differences at the chromophore binding sites that can be supposed to underlie the wide range of spectral sensitivities of arthropod photoreceptors (Stavenga and Schwemer, 1984). The binding sites are protected within the lipid bilayer, and an epitope analysis of *Cherax* rhodopsin tends to support a prediction that most monoclonal antibodies are raised to the hydrophilic loops that lie outside it (de Couet and Sigmund, 1985).

Some contradictions may be explained by a heterogeneity of arthropod opsins. O'Tousa *et al.* (1985) and Zuker *et al.* (1985) sequenced the gene for *ninaE* opsin which is expressed in photoreceptors R_{1-6} of *Drosophila*. Cowman *et al.* (1986) subsequently sequenced the gene for the opsin specifically expressed in R_8. The R_8 opsin exhibits 67% homology to that of R_{1-6}, major differences being found in the hydrophilic regions of both the cytoplasmic and luminal sides of the membrane. Similarly, molecular genetic analysis of the genes encoding the blue, red and green photopigments implicated in human colour vision shows that the hydrophilic domains are also variable (Nathans *et al.*, 1986). A great deal of diversity seems likely across the range of photopigments in arthropods.

The rhabdoms of *Cherax*, *Procambarus*, *Drosophila* and *Lucilia* are of fairly fixed volume, while those of *Leptograpsus* are of variable volume and exhibit radical turnover strategies (Section 3.2). A modest inference is that some anomalies may relate in part to species differences affecting rhodopsin chemistry; luminal oligosaccharide chains might conceivably act as signals initiating or controlling some of the events of turnover (O'Brien, 1976).

6.2 CONTROL OF BIOGENESIS

Biogenesis of the microvillar membrane demands both the synthesis of phospholipids as a bilayer by the RER, and the synthesis and insertion of opsin (or rhodopsin) into the nascent membrane. However it may be translocated, the membrane–photopigment complex must eventually be re-organized as microvilli.

For the photoreceptors of a vertebrate, Fliesler *et al.* (1985) have shown that assembly of outer segments is disrupted if glycosylation is blocked by tunicamycin. The situation in arthropod photoreceptors is clearly different. Firstly, microvillar assembly and organization do not depend upon the presence of photopigment at all. Schinz *et al.* (1982) from an elegant genetic analysis of the consequences of Vitamin A deficiency and of mutations that block the synthesis of rhodopsin in *Drosophila* to various extents found microvilli of normal shape in all states of rhodopsin deficiency; lack of photopigment was revealed by diminished populations of P-face particles in freeze-fracture preparations that could amount to total absence. Nevertheless, what happens to the turnover of such microvilli in the absence of effective transduction is not clear; signals that determine some turnover events may be missing, and non-functional microvilli may perhaps be stable after their morphogenesis during late pupal development.

Secondly, there are important effects of chromophores or their precursors. Brammer and White (1969) showed that the SER in receptors of Vitamin A-deprived mosquito larvae assumes an abnormal configuration which resembles a similar configuration (crenate SER) seen by Stowe (1980) in some SER saccules of *Leptograpsus* during rapid membrane synthesis at dusk. Paulsen and Schwemer (1983) have found that 11-*cis*-retinal stimulates the biogenesis of opsin in blowfly, and may be necessary for it to occur during normal turnover, although it has been noted above that it may not be required for the initial morphogenesis of microvilli.

6.3   AUTORADIOGRAPHIC STUDIES OF BIOGENESIS

Several pioneer studies of biogenesis employed labelling by ^{3}H-leucine as a marker for transductive membrane synthesis or renewal (Hafner and Bok, 1977; Krauhs et al., 1978; Pepe and Baumann, 1972; Perrelet, 1972; Tuurala and Lehtinen, 1974). They were conducted without an understanding of the real timing of daily cycles of turnover, making interpretation difficult. Furthermore, the assumption that the major fraction of ^{3}H-leucine incorporated into rhabdoms would label rhodopsin, although probably correct, ignored events in the microvillar lumen. Leucine is such an ubiquitous component of proteins that this is a serious problem. Hafner and Bok (1977) found that multivesicular bodies (mvbs) were labelled by ^{3}H-leucine slightly ahead of the rhabdom in crayfish. This probably relates, as they suggested, to a rapid synthesis and injection of lysosomal enzymes (Section 7).

7   Breakdown of transductive membrane

7.1   MEMBRANE SHEDDING

The predominant route for the shedding of membrane by photoreceptor microvilli is by pinocytosis at their bases (Eguchi and Waterman, 1967; White, 1967b; Blest, 1978; Blest et al., 1978a, 1980a). Small, coated vesicles are pinched off at the bases of the microvilli, and assembled into multivesicular bodies (mvbs) in the receptor cytoplasm. Multivesicular bodies are membrane-bound, or composed of tight aggregates of vesicles without an enveloping membrane. Both types are found in the spider Dinopis (Blest, 1978). Blest et al. (1978b) found some evidence that the limiting membrane of mvbs is derived from saccules of smooth endoplasmic reticulum; White et al. (1980) showed definitively that vesicles enter mvbs by secondary endocytosis. Derivation of the bounding membranes directly from ER is suggested by the presence of associated acyltransferases in photoreceptors of abalone (Kataoka and Yamamoto, 1985) with an implication that envelopes may continue to expand by self-synthesis as vesicles are added to them.

A few alternative routes for the disposal of microvillar membrane in particular species are indicated by descriptive ultrastructural studies, although in some cases they have been inferred rather than proved. These routes have been reviewed and illustrated by Blest (1980).

(1) Membrane is shed from the tips of the microvilli and endocytosed by the receptors themselves. The process is illustrated for a tipulid fly (Williams and Blest, 1980) in which the delimitation of a shedding zone is achieved by

the local deletion of the microvillar cytoskeleton (Blest *et al.*, 1982a). The route has also been inferred for a pisaurid spider (Blest and Day, 1978), and may be typical of higher Diptera (Williams, 1982a; Schwemer and Henning, 1984).

(2) Membrane is shed from the tips of the microvilli and endocytosed by adjacent glial processes; this strategy is only observed in a single species of salticid spider (Blest and Maples, 1979). It offers an interesting parallel to the relationship between photoreceptor outer segments and the adjacent pigment epithelium in vertebrates, as reviewed by Bok (1985).

(3) Microvillar membrane is endocytosed directly by adjacent photoreceptors in the crab, *Leptograpsus* (Stowe, 1983b).

(4) Blest and Price (1981) describe an unusual pattern of endocytosis by intermediate segments in *Dinopis*, locked to the daily cycle of membrane turnover, but not necessarily related to it. Some part of the process, which slightly anticipates the onset of shedding at dawn, may be concerned with the internalization of haemocyanin from extracellular space.

(5) Waterman and Piekos (1983) describe the nocturnal degradation of rhabdoms in the crayfish *Procambarus*; this only minimally involves internalization of membrane, but instead deploys phagocytic haemocytes in the retina which engulf microvillar membrane and, from the evidence of acid phosphatase (AcPh) ultrastructural cytochemistry, digest it. The haemocytes undergo cyclical migration to and from the retina, passing across interstices in the basement membrane to reach it (Waterman and Piekos, 1981, 1983; Piekos and Waterman, 1983).

It should be borne in mind, however, that some classes of haemocyte play an active role in the repair of damaged neuroglia (Smith *et al.*, 1986). Rapid migration of haemocytes into the blowfly retina after mechanical damage has been observed by S. R. Shaw (personal communication).

No species confines itself to a single strategy, and greater or lesser degrees of basal pinocytosis accompany all of them with the possible exception of that seen in the salticid spider described by Blest and Maples (1979).

7.2   DEGRADATION OF TRANSDUCTIVE MEMBRANE AND ASSOCIATED
        LYSOSOMAL SYSTEMS

Whether or not rhodopsin and other integral membrane proteins are degraded after internalization or recycled is an important issue. Many studies (White, 1967b; Eguchi and Waterman, 1976; Blest *et al.*, 1978a) have disclosed a stereotyped sequence of events. Vesicles within mvbs coalesce in such a way that the resulting "combination bodies" (Hafner *et al.*, 1980) contain both vesicles and lamellae (Fig. 8). Combination bodies are converted, in

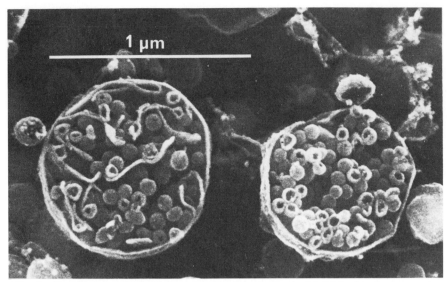

Fig. 8   Combination body (*left*) and multivesicular body (*right*) in the photoreceptor
of a crab, *Ocypode*. The fractured organelles were observed by high-resolution, field-
emission scanning electron microscopy (see Stowe *et al.*, 1986). (From Blest *et al.*,
1984b; cover illustration.)

turn, to multilamellar bodies (mlbs) which then condense to form so-called
residual bodies and lipofuscin bodies. Here, there has been a problem of
terminology. Traditionally, "residual bodies" and the lipofuscin bodies of
vertebrate neurones were thought not to be further degraded, and accumulation of the latter accompanies central neural senescence. However, Ivy *et al.*
(1984) showed that injection of either leupeptin (a cysteine protease inhibitor) or of chloroquine (a non-specific lysosomal inhibitor) into the brains of
rats caused an immediate accumulation of dense bodies in accessible neurons; this implies that they are normally degraded quite rapidly. In arthropod
receptors, both are removed. Not only can they be captured in the process of
lysis but, were they not dissipated, a photoreceptor would soon become filled
with them as a consequence of repeated cycles of turnover. Nevertheless,
ultrastructural evidence of their disposal only establishes that membrane
phospholipids are eliminated. The fate of proteins and associated chromophores is uncertain.

Indirect evidence for the degradation of rhodopsin must rely upon the following line of argument: The secondary lysosomal pathway is simple; it does
not include any ultrastructural symptoms suggestive of routes that might
segregate proteins for recycling from those destined to be degraded. Such
segregation has been determined by Geuze *et al.* (1983) for the internalization

of asialoglycoprotein by hepatocytes, proteins destined for return to the plasma membrane being isolated in a compartment designated as "CURL" which should be readily recognizable in electron micrographs. Given that rhodopsin is an integral membrane protein whose major domains are hydrophobic (O'Tousa et al., 1985; Zuker et al., 1985), it is difficult to conceive its fate were it to be isolated from the phospholipid bilayer in which it is embedded. How would it be transported in the cytosol, and, how would it be re-inserted into nascent microvillar membrane? The elaborate arrangements for the synthesis and insertion into membrane of a vertebrate rhodopsin disclosed by Goldman and Blobel (1981) make the problem all the more intractable.

Some direct evidence that rhodopsin is degraded in the lysosomal pathway was provided by Eguchi and Waterman (1976), who observed that the density of P-face particles in freeze-fracture preparations decreases from the high densities noted in microvillar membranes and in those of vesicles within mvbs, to low densities in the membranes of mlbs. The conclusion from this study is prejudiced by the small number of organelles sampled, and by technical difficulties imposed by the changing membrane conformations throughout the ultrastructural sequence of events.

The degradation of secondary lysosomes demands specific enzymes. Daily cycles of turnover that internalize massive amounts of membrane at dawn imply an equivalently massive injection of enzymes into the secondary lysosomal pathways. Ultrastructural cytochemistry of acid phosphatase (AcPh) has been employed in several studies because it is both simple and reliable. Some authors have failed to realize that AcPh is a general marker for lysosomal organelles, not because it degrades key proteins but because it is concerned with the deletion of the mannose-6-phosphate recognition marker that is believed to direct all lysosomal enzymes to their destinations (Sly et al., 1981).

Tentative flow diagrams for the primary and secondary lysosomal compartments in Dinopis and Leptograpsus are given in a review by Blest et al. (1984b). A diagram for Leptograpsus is shown in Figure 9. Blest et al. (1979) examined the distribution of AcPh by ultrastructural cytochemistry in the receptors of Dinopis following internalization of membrane at dawn. Reactions were satisfactorily displayed in the very large residual bodies typical of that species, and in saccular sheets of rough endoplasmic reticulum in the process of being incorporated into them. The Golgi is unobtrusive in Dinopis and was not labelled.

A more substantial picture was obtained by Blest et al. (1980a) for the receptors of the crab Leptograpsus. Reactions for AcPh first appear in mvbs, just before vesicles start to transform to lamellae, and strong reactions are found in mlbs (Fig. 10). The Golgi is seldom labelled, but vesicles considered

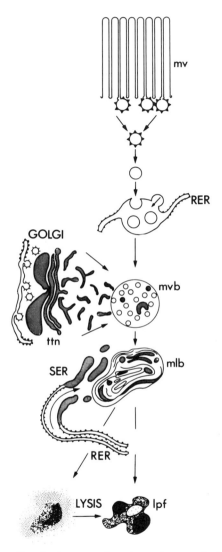

Fig. 9   Flow diagram for the lysosomal pathway in a crab, *Leptograpsus*. Coated vesicles are pinched off from the bases of the microvilli (mv) and enter saccules derived from the endoplasmic reticulum (RER) by secondary endocytosis. They form multivesicular bodies (mvb) which transform to multilamellar bodies (mlb). Lysosomal enzymes are injected into these secondary lysosomes via a transtubular network (ttn) of the Golgi apparatus, and from smooth saccules (SER) derived from the endoplasmic reticulum. Early secondary lysosomes are either lysed directly, or condense to form lipofuscin bodies (lpf) which are lysed in turn. (From Blest *et al.*, 1984.)

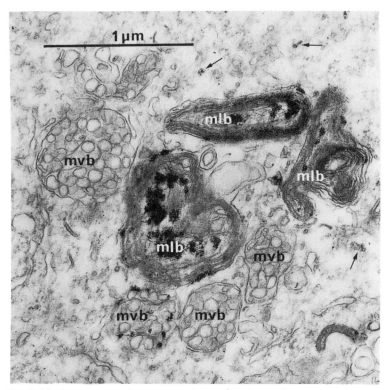

Fig. 10 Acid phosphatase ultrastructural cytochemistry of secondary lysosomes of *Leptograpsus*. Multivesicular bodies (mvbs) are either not labelled or, as in the case of the mvb at bottom left, show a weak punctate labelling as they transform to combination bodies. Multilamellar bodies (mlbs) are strongly labelled. Feeble responses are sometimes seen in vesicles and saccules derived from the Golgi or from the endoplasmic reticulum (*arrowed*), but may be artifactual. All labelling is absent in controls in which acid phosphatase is inhibited by fluoride, or from which the substrate, β-glycerophosphate is omitted. Activity was revealed by a standard lead-capture method. (× 47000.) For further discussion see Blest *et al.* (1980a).

to be derived from it become reactive as they fuse with mlbs. A contribution from the ER was also seen to be labelled as it fused with lysing mlbs. Aryl sulphatase has not been demonstrated in any arthropod lysosomal system, although it is present in that of abalone photoreceptors (Kataoka and Yamamoto, 1985). The consistent failure of the Golgi to be labelled in arthropod systems is puzzling, for it is readily labelled by AcPh ultrastructural cytochemistry in both abalone and a sabellid worm (Kataoka and Yamamoto, 1985; Eakin and Brandenburger, 1985; Brandenburger and Eakin, 1985), just as it is in locust central neurones (Lane and Swales, 1976).

Populations of secondary lysosomes follow the trends that may be predicted from the temporal patterns of internalization of membrane. They have
been meticulously quantified in crayfish by Hafner *et al.* (1980). They cannot
be used as absolute measures of rates of membrane turnover because their
persistence and, presumably, rates of transition between subcategories, will
depend upon when and in what quantity lysosomal enzymes are introduced
into them, and how fast lysis proceeds. De Couet and Blest (1982) attempted
to relate titres of crab AcPh to the daily pattern of membrane turnover in
Leptograpsus, using whole retinae. Correlations were blurred by at least two
factors, so that the correspondence observed was less than might have been
expected; firstly, there is a substantial turnover of both lipoprotein bodies
and mitochondria in the receptors, whose interaction with the daily cycle is
unknown. Secondly, assays for AcPh at pH 5·0 are now known to include the
acid tail of reactions mediated by a cytosolic inositol polyphosphatase and a
phosphoprotein phosphatase, both with neutral pH optima (Trowell, 1984
and in press). Nevertheless, levels of AcPh rise significantly around dawn,
when membrane is in the process of being massively degraded.

Occasionally, authors continue to speculate that mvbs may carry newly
synthesized plasma membrane to nascent microvilli (Brammer *et al.*, 1978;
Meyer-Rochow, 1982). Mvbs, however, seem merely to fulfil the catabolic
roles that have been well-established for vertebrate cells and which are confirmed for invertebrate receptors by ultrastructural cytochemistry.

7.3   TURNOVER OF MICROVILLAR CONTENTS

Radical turnover strategies, especially those of locust, *Leptograpsus* and
Dinopis which can involve the complete demolition and reassembly of a rhabdom, imply that microvillar contents must be either destroyed or recycled as
well. Nothing is known about such cytosolic processes. Possible events that
implicate the cytoskeleton will be considered below, in the discussion of
models for the regulation of turnover and for the assembly of microvilli
(Section 8.3).

7.4   SELECTIVE INTERNALIZATION OF INTEGRAL MEMBRANE PROTEINS

Schwemer (1984) used microspectrophotometry to show that metarhodopsin
is internalized preferentially during turnover by photoreceptors in the
blowfly. It is not known whether similar selectivity is also exhibited by other
arthropods with relatively stable, fixed-volume rhabdoms. A comprehensive
flow diagram for the turnover of visual pigment in the blowfly *Calliphora*,

which takes into account the photopigment transitions, is given by Schwemer (1986) and is shown in Figure 11.

7.5 PHOTORECEPTOR MEMBRANE BREAKDOWN IN *LIMULUS*

The fate of internalized membrane in *Limulus* is still unknown; a number of studies reviewed by Barlow and Chamberlain (1980) and Chamberlain and Barlow (1984) suggest a picture different from that of other arthropods. At dawn, rhabdomeral organization disintegrates, with the formation of membrane whorls from which mvbs appear to be secondarily derived. Throughout the subsequent day, mvbs successively reorganize to yield combination bodies and mlbs. Uniquely, normal shedding at dawn does not proceed by pinocytosis from the bases of the microvilli; the mode of transformation of primary membrane whorls to mvbs in the cytoplasm has not yet been elucidated. Furthermore, following dawn breakdown, there is a rapid restitution of microvillar membrane area, to a value that seems relatively stable for both night and day.

Unfortunately, these studies do not indicate how new membrane is synthesized and assembled, nor do they utilize ultrastructural cytochemistry to analyse lysosomal pathways. Stowe (1981) has noted that the time resolutions of turnover events offered by these authors seem incompatible with the small number of animals sampled in their earlier studies.

An implication that transductive membrane may be segregated from the rhabdoms at dawn and returned to them without being degraded has an interesting parallel in the crab *Callinectes*. Toh and Waterman (1982) suggest that shed membrane is not degraded and is available for reincorporation into the rhabdom.

7.6 MEMBRANE TRAFFIC AND SOME CONSEQUENCES

Flow of membrane to and from microvillar bases is implicit in the events of turnover. It provokes a number of general questions.

(1) Given that SER cisternae are in continuity with microvillar bases in higher Diptera, and may replace microvillar membrane as it is shed at their tips (Section 5.1), it is reasonable to ask what their various functions may be. Walz (1982a–c), in an important series of papers, has shown that such subrhabdomeric cisternae sequester calcium. It is released in response to the second messenger, 1,4,5-inositol trisphosphate (Brown *et al.*, 1984; Fein *et al.*, 1984). Do all such SER cisternae sequester calcium? Are they different from the cisternae that are presumed to contribute to regenerating microvilli?

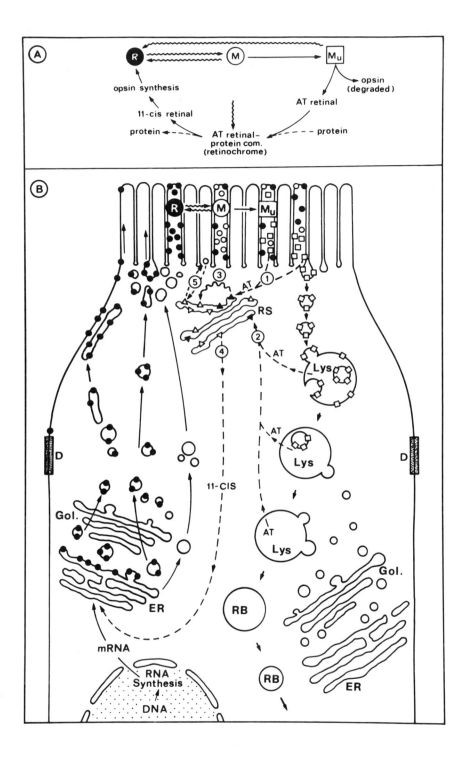

In bee, Skalska-Rakowska and Baumgartner (1985) have provided good evidence that subrhabdomeric cisternae are in total continuity as a single compartment within a photoreceptor.

For blowfly, Walz (1982c) has demonstrated a patchy distribution of sequestered calcium in subrhabdomeric cisternae, whether there is only one compartment for a single photoreceptor or not. Two alternative models seem to be implied: (a) all SER cisternal membrane may contribute to the regeneration of microvilli, irrespective of local roles in the sequestration and release of calcium; (b) Only regions of SER cisternal membrane that are not concerned with calcium sequestration contribute to microvillar regeneration.

(2) If membrane flow from the ER is the mechanism for microvillar renewal (as proposed by Itaya, 1976 and Stowe, 1980), what are its implications for proximal regions of the microvilli? Hamdorf and Kirschfeld (1980) infer that there may be few, or even just one, ion channel per microvillus and that channels may be located at microvillar bases. The bases of microvilli in the blowflies that they studied are, given existing models of renewal, in a continuous state of flux, presenting some obvious problems.

Fig. 11 Schematic representation of visual pigment turnover. The diagram combines data obtained by microspectrophotometry from blowflies, and by ultrastructural cytochemistry from crabs. Hypothetical pathways are indicated by broken lines. (A) Rhodopsin ($\bullet$) is converted by light into metarhodopsin, M ($\circ$), which in turn is converted into unstable metarhodopsin M_u ($\square$). Both forms of metarhodopsin can be reconverted to R by light. All-*trans*-retinal (AT) is bound to a protein to form a retinal–protein complex. All-*trans*-retinal is photoisomerized to yield 11-*cis*-retinal, which induces opsin synthesis. *Wavy lines* indicate reactions initiated by light; solid lines indicate reactions that proceed in darkness; broken lines, reactions or pathways that are currently hypothetical. (B) A schematic representation of the visual pigment cycle, using the conventions of (A). R ($\bullet$) is converted to M ($\circ$), which transforms to M_u ($\square$). M_u is transported to the bases of the microvilli, removed by endocytosis, and degraded by a lysosomal system (*right*). It is assumed that all-*trans*-retinal is released at some stage from M_u and recycled, possible sites of release being the microvilli (1), or the secondary lysosomes (2). The AT released may be bound to a protein to form a photopigment, possibly a retinochrome; however, the complex family of retinal-binding proteins responsible for chromophore transport in the retinae of vertebrates (Bok, 1985) allow more elaborate alternatives. A retinochrome ($\blacktriangle$) would be converted by light (3) to a metaretinochrome ($\triangle$) containing the chromophore in the 11-*cis* form, which would be transported to the RER (4) to serve as a prerequisite for the initiation of opsin synthesis. The synthetic pathway is indicated on the left-hand side. Opsin is synthesized in the RER, and delivered via a pathway that may include the Golgi to the microvillar bases, or to the flanking plasma membranes. Desmosomes (D) represent a diffusion barrier for visual pigment molecules in the plasmalemma. ER, endoplasmic reticulum; RS, smooth endoplasmic reticulum associated with microvillar bases; Gol, Golgi complexes; Lys, secondary lysosomes; RB, residual bodies. (Modified from Schwemer, 1986.)

8   Some models for turnover events

8.1   MODELS FOR THE INTERNALIZATION OF MEMBRANE

It is reasonable to ask whether the internalization and presumptive degradation of transductive membrane by arthropod photoreceptors resemble more
familiar processes in other types of cell. The intensively studied events of
receptor-mediated endocytosis (Goldstein *et al.*, 1979) appear promising as a
model: coated pits selectively internalize particular membrane components
(Bretscher, 1976). The complicated alternative fates of receptor-ligand complexes after internalization are categorized by Wileman *et al.* (1986). There is
no suggestion that the endocytosis of photopigment involves binding to a
ligand, but the selective internalization of metarhodopsin reported by
Schwemer (1984) implies that the processes could be similar. It would be
interesting to know whether *any* metarhodopsin molecule is a candidate for
degradation, or whether a subpopulation of chemically or sterically damaged
metarhodopsins is sequestered because functionally it cannot participate in
the regeneration of rhodopsin.

Coated pits in flat regions of the photoreceptor plasmalemmae of *Lucilia*
have presumptive microfilaments radiating from them (Blest *et al.*, 1984a);
this resembled similar dispositions on the coated pits of vertebrate cells
(Inoue, 1981; Aggeler and Werb, 1982). What may happen within microvilli
is a more difficult matter, and cannot realistically be discussed because of the
still ambiguous nature of the microvillar cytoskeleton. In *Drosophila*, *Lucilia*
and *Cherax* (Blest *et al.*, 1984a; de Couet *et al.*, 1984) and also in squid
(Saibil, 1982) conventional transmission electron microscopy only reveals a
single, axial filament, bound to the plasmalemmae by side-arms. There are
three problems, and these are listed below.

Although the reason is not apparent, translational diffusion of rhodopsin
molecules within the phospholipid bilayer is highly constrained to the point
where, in crayfish, it has not been possible to demonstrate diffusion at all
(Goldsmith and Wehner, 1977). Since coated pits and pinocytotic vesicles
form at the bases of microvilli, the constraint must be lifted for molecules
selected for disposal should Schwemer's paradigm for the case of the blowfly
prove true for all arthropods.

The nature of the axial filament within microvilli is still uncertain. In crayfish, the cytoskeleton appears to be composed of F-actin, but the appearance
of a single filament in *Cherax* must be taken in conjunction with the finding
of several microfilaments within a microvillus of the crayfish *Procambarus* by
J. Usukura (unpublished results), using both freeze-substitution and deep
etching techniques, and the finding of several microfilaments within each of

the interdigitated microvilli of some nocturnal spiders by Blest and Sigmund (1985). Furthermore, in higher Diptera, axial microvillar filaments appear thick in transmission electron micrographs (Blest *et al.*, 1982a, b) and run parallel to and possibly in continuity with robust filaments in the extra-rhabdomeral cytoplasm (Blest *et al.*, 1984a). It would be reckless to describe such robust elements as "intermediate filaments" because, as is well known, the axons of insects are devoid of neurofilaments (e.g. Benshalom and Reese, 1985). Evidence from cultured, non-neural cells for the presence of inter-mediate filaments in *Drosophila* tissues has only been provided by Walter and Biessman (1984).

Nevertheless, a molecular genetic analysis of the *Drosophila* phototransduc-tion mutant, *ninaC*, established an abnormality in the microvillar cytoskeleton as the primary defect. The *ninaC* gene encodes two photoreceptor-specific polypeptides with molecular weights of 130,000 and 170,000 KD both of which cross-react weakly with antibodies to vertebrate neurofilaments (W. Pak, personal communication). Thus, the nature of the microvillar cyto-skeleton is still unknown, with the additional *caveat* that cytoskeletal organ-ization throughout the arthropods may prove to be diverse, and to have a variety of roles. These ambiguities are important, because their resolution will necessarily determine how the mechanism of turnover of rhabdomeral membrane and of its constituent proteins comes to be interpreted in the future.

Whatever its nature, much may depend upon whether the microvillar cytoskeleton is static or dynamic. It is possible to conceive that "tread-milling" of a dynamic F-actin cytoskeleton bonded appropriately to integral membrane proteins might be employed to sweep molecules destined for degradation to the microvillar bases (Blest *et al.*, 1984b).

8.2   MODELS FOR EXTRACELLULAR SHEDDING

Abscission of membrane from the tips of microvilli (Blest and Maples, 1979; Blest *et al.*, 1982b; Williams and Blest, 1980) demands that a local trans-formation must take place that allows it to happen. In photoreceptors of tipulid flies, there is a discrete "shedding zone" occupying the distal region of a rhabdomere; it appears relatively electron-lucent in micrographs. Micro-villi in the shedding zone are empty bags of membrane; a cytoskeleton is ab-sent, but the distribution of P-face particles in the microvillar membranes of the shedding zones does not differ from that of the stable, basal regions (Blest *et al.*, 1982b).

A two-stage process is implied; deletion of a microvillar cytoskeleton that precedes shedding, and the process of shedding itself. The former step requires a very exact control of local cytoskeletal domains, the latter a signal to precipitate the dramatic shedding event illustrated by Williams and Blest (1980), during which the whole of a shedding zone vesiculates as it is abscissed, and is afterwards endocytosed from the extracellular space by the photoreceptors.

8.3   MODELS FOR THE ASSEMBLY OF NEW MICROVILLI

Although it is attractive to suggest that microvilli are erected by the polymer-ization of actin and the bundling of microfilaments, as has been suggested for microvilli of the brush-border by Tilney (1979), there is little to suggest that this model is likely. The conformational changes observed by Stowe (1980) during replacement of the day rhabdom in *Leptograpsus* are not of the right kind to support it.

The formation of extracellular bridges between microvilli might also sus-tain their erection, but it is not certain that the microvilli of all rhabdomeres possess them. They have so far only been seen in the rhabdomeres of blowfly (Blest *et al.*, 1982a), mosquito larva (White, 1967a), squid (Saibil, 1982) and bee (Blest, unpublished observations), and appear to somewhat resemble septate junctions.

Is it necessary to assume a cytoskeletal mechanism for the generation of microvilli at all? An alternative possibility might be that the tubular con-formation is thermodynamically best once all the integral membrane proteins are in place. Against this interpretation are observations which suggest that microvilli of normal dimensions are formed irrespective of whether or not they contain a normal complement of rhodopsin (Schinz *et al.*, 1982), and that in the *Drosophila* mutant *ninaC* an abnormality of the cytoskeleton is accompanied by distortion of the microvilli (W. Pak, personal communica-tion).

An interesting comparison is afforded by the generation of large masses of crystalloid smooth endoplasmic reticulum in cultured hamster UT-1 cells fol-lowing treatment with compactin (Chin *et al.*, 1982; Anderson *et al.*, 1985). A rate-limiting enzyme in the pathway for cholesterol synthesis, 3-hydroxy-3-methylglutaryl Coenzyme A reductase, is an integral membrane protein. Induced to accumulate pharmacologically, the endoplasmic membranes bearing it form regular, crystalloid masses of intracellular tubules whose dimensions are similar to those of rhabdomeral microvilli. These bear a strik-ing resemblance to the architecture of rhabdoms, although it must be noted that, in relation to plasmalemmal microvilli, endoplasmic tubules are "inside out". Nevertheless, there is no suggestion, so far, that a cytoskeleton plays any part in their assembly, or that the tubules contain one (Anderson *et al.*, 1985).

The question of how photoreceptor microvilli are assembled rapidly and in the right place is thus at present completely open.

9 Light-induced retinal damage

9.1 SOME DEFINITIONS

Turnover of rhabdomeral membrane is normal; although in many species catastrophic breakdown is triggered at dawn by light (Section 3.1), the phenomena observed should not be regarded as pathological. This stricture applies, for example, to the radical disorganization of the locust rhabdom as it diminishes at dawn (Horridge *et al.*, 1981; Williams, 1982b). Not all rhabdoms that suffer catastrophic reduction at dawn exhibit disorganization. Those of *Leptograpsus*, for example, are reduced in a more orderly manner (Blest *et al.*, 1980a; Stowe 1983a, b).

Pathological breakdown occurs in at least three circumstances:

(1) A "night" rhabdom is illuminated at modest levels soon after it has been assembled, and long before natural dawn (Blest, 1980 in *Dinopis*; Horridge *et al.*, 1981; Williams, 1982b in locust);

(2) Rhabdoms receive levels of illuminance grossly in excess of those offered by the natural habitats of particular species (Loew, 1976, 1980; Shelton *et al.*, 1985 in the lobster *Nephrops*; Meyer-Rochow and Eguchi, 1984; Meyer-Rochow and Tiang, 1979 in various crustacea).

(3) Rhabdoms are exposed to experimental levels of light greater than those found in nature (other than by looking directly at the sun).

Meyer-Rochow and Tiang (1979, 1982) showed that for an antarctic amphipod living at low temperatures, some of the effects of raising ambient temperatures quite modestly seem to mimic the appearances of pathological light-induced damage, a consequence, perhaps of a low phospholipid fluidity adapted to cold habitats.

9.2 CONSTRAINTS ON INTERPRETATION

The disorganized photoreceptor membranes that result from light damage are difficult to interpret because it is not clear in all cases to what extent fixation artifacts have contributed to their appearance. White and Michaud (1981) and Williams (1980a) found that divalent cations added to glutaraldehyde-fixed rhabdoms caused the formation of membrane whorls during subsequent osmication, although the mechanism for this is not understood. Williams (1980a) suggested that regions of rhabdom that are about to shed membrane are most vulnerable to such damage; rhabdomeres of blowfly shed membrane at the microvillar tips, at their bases, and also appear to dispose of entire microvilli at their sides during shedding (Williams, 1982a; Schwemer, 1986). It is these regions that generate membrane whorls in response to divalent cations. Similarly, in tipulid flies, persistence of Mg^{2+} throughout primary and secondary-fixation causes the whole of the shedding zone to collapse, and whorls to form at the bases of the microvilli, where shedding by pinocytosis also takes place (Blest et al., 1982b). Some implications of the artefactual generation of such "myelin figures" in many of the earlier ultrastructural studies of compound eyes should be mentioned. Various earlier studies of marine animals employed fixation media made up in sea-water, which has a high calcium content. In some laboratories, calcium was added to fixatives. It was believed that its presence improved the preservation of membranes. No workers before Blest et al. (1982a, b) routinely chelated divalent cations from fixation media and/or osmication media with EGTA or EDTA. Finally, the high intracellular Ca^{2+} fluxes generated during strong light adaptation may also be significant in some ultrastructural contexts.

Thus, the reality of membrane configurations that have been found to follow light-induced rhabdomeral damage should always be questioned. Disordered membrane has only once been proved to be non-artefactual by the use of freeze-substitution fixation in an exemplary study by Meinecke and Langer (1984) on the compound eye of the noctuid moth Spodoptera.

9.3   THE PHENOMENOLOGY OF LIGHT-INDUCED DAMAGE

The most careful and intensive study of light-induced damage in the rhabdoms of an invertebrate has been conducted by Shelton et al. (1985) on the burrowing lobster Nephrops, following observations by Loew (1976, 1980), and by Nilsson and Lindstrom (1983) on another crustacean, Cirolanus. Nephrops exhibits a similar pattern of daily changes of rhabdom volume to those described in the crab Grapsus (Nässel and Waterman, 1979), although they are less marked, amounting to no more than a fourfold increase in volume at night. This daily cycle is observed in Nephrops retrieved from its normal (and at all times dimly illuminated) habitat at a depth of about 25 m or more. Nephrops raised to the surface and exposed to daytime ambient illumination for periods as short as 9 minutes undergo a profound and irreversible disorganization of their rhabdoms, the surface light flux amounting to circa 228×10^{18} photons/m^2/second. Animals exposed to a similar light flux (223×10^{18} photons/m^2/second) for 15 minutes already exhibit changes to the photoreceptors. These include the disappearance of clear boundaries between the rhabdomeral contributions from adjacent receptors, and the arrival of haemocytes in the retina. These trends are extended over a subsequent 2-hour period; the rhabdoms become unequivocally disorganized, with the formation of membrane whorls from their microvilli. Six hours later, a more massive invasion of the retina by haemocytes which have migrated across the basement membrane of the eye is evident. It is likely that haemocytes have a phagocytotic role (Section 7.1; Waterman and Piekos, 1981, 1983).

The sensitivity of the retina of *Nephrops* to levels of illumination above those of their natural habitat (thought to be around 0.14×10^{18} photons/m/second—the level used in experimental holding tanks) is well illustrated by the effect of exposing animals thus manipulated and in a dark-adapted condition, to a daylight level of circa 108×10^{18} photons/m^2/second for only 15 seconds. A fortnight after this brief exposure, the retina is seen to be massively damaged. The retinae of light-adapted animals, in which protection is afforded by the migration of screening pigment, are not affected by such short exposures, but are equivalently damaged following a 5-minute exposure.

Somewhat similar, but unquantified effects of night exposure of the retina of *Dinopis* to low levels of illumination are noted by Blest (1980). Why do these crustacean rhabdoms deteriorate after such short exposures to light? Some models for light-induced damage will be considered below. In *Procambarus*, steady state conversion of photopigment to 20% rhodopsin and 80% metarhodopsin by orange-adapting light produces marked changes in the morphology of the rhabdoms, in the distribution of accessory screening pigment, and causes the microvilli to vesiculate (Cronin and Goldsmith, 1982, 1984). The degree of microvillar damage seems to relate directly to metarhodopsin content. Cronin and Goldsmith (1984) hypothesize that recovery of rhodopsin requires the synthesis and assembly of new membrane. Cummins and Goldsmith (1986) provide indirect electrophysiological evidence that intracellular Ca^{2+} levels are raised in damaged cells.

9.4   MODELS OF LIGHT-INDUCED DAMAGE

A number of models seem possible as explanations of light-induced retinal damage, although they do not seem equally plausible for all of the conditions categorized in Section 9.1. Firstly, Crouch (1980) has made a good case for the idea that damage by superoxide radicals underlies the progressive failure of isolated bovine rod outer segments to regenerate rhodopsin, after bleaching by strong illumination followed by the supply of 11-*cis* retinal to the suspension. The effect is significantly diminished if superoxide scavengers are present during illumination. Opsin seems unlikely to be especially vulnerable to superoxide radicals, but unsaturated phospholipids are. Regeneration of rhodopsin from opsin after bleaching requires the presence of a surrounding phospholipid *milieu* (Henselman and Cusanovich, 1974) whose oxidative modification could be prejudicial.

It would be interesting to discover the distribution of superoxide dismutases in the retinae of invertebrate species, and to ascertain whether they are present in large amounts in animals whose normal habitats expose them to very high levels of illumination.

An alternative, and attractive proposal has been made by Kirschfeld (1982) and Zhu and Kirschfeld (1984) who discussed the possible roles of photostable carotenoid pigments in protecting photoreceptors against photosensitized oxidative damage. They argue that components of mitochondria and other organelles, such as porphyrins, may render photoreceptors exceptionally vulnerable to high light intensities. In *Musca*, two-thirds of all R_7 photoreceptors contain a photostable yellow pigment (R_{7y}), whilst one-third do not (R_{7p}). When R_{7y} are illuminated for circa 10 minutes with ultraviolet light, their rhodopsin concentrations remain constant, whereas those of R_{7p} decrease continuously and linearly with time. Similar protective effects of carotenoids in bacteria and plants have been documented by Krinski (1979).

Again, a study of the distribution of protective pigments among different species would be informative. It is possible that they have been acquired in parallel with the evolution of "stable" rhabdoms, and that contemporary "variable size" rhabdoms do not possess them. Cases such as that of *Nephrops*, *Dinopis* and locusts are unlikely to be explained in terms of oxidative effects because the light levels and durations of exposure to them that produce massive damage are simply too low and too short for such an explanation to be plausible. What are the alternatives?

9.4.1   *Osmotic damage*

There is indirect evidence that the processes of light/dark adaptation in some arthropods may be accompanied by marked changes of cytosolic osmolarity (Blest, unpublished observations). Briefly, in both locusts and tettigoniid orthoptera, a primary fixative of given osmolarity will, for electron microscopy, preserve receptors in either the day or night condition undistorted, but seldom both. It could be argued that excessive illumination exaggerates osmotic responses, which alone cause disintegration of the rhabdoms.

Against this interpretation, Winterhage and Stieve (1982) have shown that the rhabdoms of crayfish are remarkably resistant to osmotic damage. Cytoskeletal degradation might, however, render microvilli more vulnerable to osmotic damage, following an example discussed by Wu *et al.* (1982).

9.4.2 *Phosphoinositide metabolism*

The release of calcium from store in many types of cell (Authi and Crawford, 1985), including arthropod photoreceptors (Brown *et al.*, 1984; Fein *et al.*, 1984) is triggered by the second messenger 1,4,5 inositol trisphosphate (InsP$_3$) (Berridge, 1984; Nishizuka, 1984), derived from the inner leaflet of the plasma membrane bilayer. What would happen if intense and sustained illumination provoked its continuous release?

An appropriate model is provided by the crenation of human erythrocytes *in vitro* (Ferrell and Huestis, 1984; Ferrell *et al.* 1985). With reasonable experimental and quantitative evidence, these authors show that crenation can be explained by an asymmetric depletion of the bilayer, with phospholipids being preferentially lost from the inner (cytoplasmic) leaflet. Substantial overproduction of InsP$_3$ by an overstimulated photoreceptor might produce an analogous result. Although such a model is attractive, it will need to be realistically evaluated in terms of the inter-relationship between phosphoinositide metabolism and calcium regulation, whose current and complex state is reviewed by Sekar and Hokin (1986).

9.4.3 *Calcium exchanges and degradative pathways*

The high calcium fluxes experienced by photoreceptors during light/dark adaptation have already been noted (Section L). Lo *et al.* (1980) argued that all the calcium that accumulated in *Drosophila* photoreceptors during reversible poisoning by the metabolic inhibitor, 2,4-dinitrophenol, was released from intracellular stores; cytosolic levels of Ca^{2+} could reach 2·5 mM after sustained poisoning. Possibly, an overproduction of InsP$_3$ in response to intense illumination could also elicit a total release of stored Ca^{2+} into the cytosol. What consequences might ensue? Much would depend upon the role of the calmodulin associated with rhabdoms (de Couet *et al.*, 1986): how effectively might it buffer Ca^{2+} fluxes, and does it, in fact, exert such a buffering role *in vivo*?

Large fluxes of Ca^{2+}, inadequately buffered, might influence an actin cytoskeleton in three alternative ways: From the model of the vertebrate brush-border (Mooseker, 1983), Ca^{2+} fluxes might be expected to disassemble microvillar cytoskeletons, *via* Ca^{2+}-dependent actin-regulating proteins. Alternatively, if such regulating proteins are not present in rhabdomeral microvilli, it is possible that Ca^{2+} fluxes might promote polymerization of actin, in terms of the model of actin regulation proposed by Europe-Finner and Newell (1986) for the generation of pseudopodia by the slime-mould *Dictyostelium* in response to cyclic AMP as an extracellular signal. In that case, amoebae of *Dictyostelium*, permeabilized by saponin, exhibited actin polymerization in response either to Ca^{2+}, or to a low level of inositol 1,4,5 trisphosphate which was presumed to release Ca^{2+} from intracellular stores. The Ca^{2+} fluxes can also be supposed to activate Ca^{2+}-activated neutral cysteine proteases, for whose presence in arthropod photoreceptors there is strong indirect evidence (Blest *et al.*, 1982a, b; Blest *et al.*, 1984a; de Couet *et al.*, 1984). Immediately, proteolysis would degrade microvillar cytoskeletons, but such an effect of Ca^{2+} fluxes could be inhibited by associated cystatins. Cytoskeletal proteolysis would be unlikely to generate anything more than the microvillar distortions observed in the cytoskeleton-depleted shedding zone of the rhabdomeres of tipulid flies (Blest *et al.*, 1982), but could, as has been noted, render a rhabdomere more vulnerable to osmotic shock.

A related pathway is suggested by recent work on mammalian blood platelets (Ruggiero and Lapetina, 1985). Thrombin binds to platelets and induces their activation. It potentiates a Ca^{2+}-dependent protease that degrades cytosketal proteins, and, especially, talin (O'Halloran *et al.*, 1985). It also induces the proteolytic activation of phospholipase C (which releases $InsP_3$ from the inner leaflet of the bilayer) *via* a Ca^{2+}-activated, leupeptin-sensitive protease (Ruggiero and Lapetina, 1985). An analogous system could plausibly underlie the mechanism of light damage to both arthropod and vertebrate retinae.

Any model that invokes Ca^{2+} fluxes must encompass the probability that calcium may be unevenly distributed within a photoreceptor, and that transitory local concentrations in the cytosol may be high (Harary and Brown, 1984).

10   Retinal damage induced by prolonged darkness

Normal cycling of phototransductive membrane requires a daily cycle of illuminance. Two studies have addressed the consequences of continuous, prolonged darkness on retinae with a substantial daily pattern of cyclical membrane exchange. Eguchi and Waterman (1979) and Hafner *et al.* (1982) examined crayfish receptors, and Bloom and Atwood (1981) studied those of locust. In both cases, light deprivation extended for periods of weeks and produced various degrees of rhabdomeral disorganization. The latter authors found that restoration of a normal light cycle for 14 days after 8 weeks of light deprivation allowed the reassembly of normal microvilli, but not repair of the pathological fusion between adjacent rhabdomeres of a compound rhabdom that deprivation induced. All the studies concluded that the effects they observed were probably the consequences of a failure of control mechanisms for turnover which are dependent upon light.

11 Discussion and conclusions

The turnover of the phototransductive membrane of arthropods is critically different from that of vertebrate photoreceptors in one important respect: in vertebrate photoreceptors, turnover routines are slow, and are not manipulated topographically in terms of diurnally varying optical goals. In arthropods, the pattern of membrane turnover is inseparable from its optical consequences. Although turnover of specialized transductive membrane may be assumed to follow pathways that have been well-established for vertebrate cells in culture, major problems concerning the maintenance of precise cell-surface architectures which must satisfy the stringent demands of day and of night vision are almost unique to compound eyes. The daily repetitive re-modelling of rhabdoms, in some long-lived arthropods over periods of years, has been described as "recurrent morphogenesis" (Blest *et al.* 1984b). Traffic of plasma membrane precursors to a rhabdom, and the pathways for the internalization and degradation of components during the catabolic phase of turnover must be very precisely regulated. Such regulation must ensure not only that enzymes of the transduction cascade are accurately localized, but that they are synthesized in appropriate amounts.

The matter of how future research on turnover routines should be conducted is, obviously, difficult. There is no well-established, traditional base for the study of arthropod cells in culture, and in any case, mature photoreceptors cannot be cultured. The amounts of arthropod photoreceptors available for study will always be limited. Nevertheless, the problems of intracellular traffic and regulation that they present are important, and dramatically exemplify many of the key problems of modern cell biology. Accurate image analysis depends upon the consistency of the responses of the individual photoreceptors that constitute a retinal mosaic. Much variability of the responses of individual photoreceptors would degrade the quality of an image which they collectively abstract. How then is the architecture of an individual photoreceptor precisely determined? Any answer must encompass the control of the transcription and translation of enzymes of the transduction cascade, their transport to and concentration at appropriate sites of action, and mechanisms that ensure that adequate precision is guaranteed for both "day" and "night" steady states.

Descriptive ultrastructural studies are unlikely to yield much more significant information, although novel, idiosyncratic pathways for the disposal and degradation of rhabdomeral membrane may still await discovery. The molecular genetic regulation of turnover, with particular reference to quantitative interactions, is one of the most attractive foci for future research.

The relationship between normal turnover phenomena and light-induced damage, currently not satisfactorily resolved even at a descriptive level, is another promising area, and especially so now that models for the membrane events have recently become available and have reached a certain level of sophistication.

Similar considerations apply to models for the translocation of nascent membrane. Attention has primarily been paid to the pathway for rhodopsin, but models derived from vertebrate cells indicate that the fate of lipids may be unexpectedly complex. Phospholipids and sterols may be transferred to the plasma membrane by distinct and quantitatively independent routes (Kaplan and Simoni, 1985a, b), opening up a possibility that the conformational changes associated with the assembly of microvilli are thermodynamically assured by late changes to the composition of the presumptive microvillar membranes.

The complicated transfers of organelles within arthropod photoreceptors also demand attention. How do they relate to, and to what extent are they determined by recently developed models for the roles of microtubules and associated proteins in the process of fast axonal transport (Schnapp et al., 1985; Vale et al. 1985)? Here, some distinction must ultimately be made between fast axonal transport through the receptive segment, the presumably directed traffic of plasma membrane precursors and their catabolic products, and the fast, directed movements of intracellular granules of screening pigment that accompany dark and light adaptation in many insect receptors. The latter migrations may prove merely to be specialized, radial versions of fast axonal transport.

It is evident that membrane turnover and allied phenomena in the photoreceptors of invertebrates have been studied fragmentarily in too many different preparations. As a preliminary, the approach has been justified by the diversity of turnover stratagems and by logistic problems. Hopefully, in the future, more sophisticated research will utilize no more than a few preparations, each accessible to a greater range of techniques.

Acknowledgements

I am grateful to the colleagues who have participated in our research for numerous discussions and much helpful advice: Dr H. G. de Couet, Dr S. B.

Laughlin, Dr S. J. Stowe, Dr S. C. Trowell and Dr D. S. Williams. None should be held responsible for any particular stance that I have adopted in this review. Throughout our work, we have received magnificent support from Mr George Weston and the staff of the Australian National University Transmission Electron Microscope Unit. We are also indebted to Dr K. Hanada, the Taisho Pharmaceutical Company, Tokyo, for generously supplying inhibitors of cysteine proteases since 1981. A succession of technical assistants has been responsible for preparing thin sections for electron microscopy: Ling Kao, Karen Powell, Joanne Maples, Dean Price, Claudia Sigmund and Margrit Carter. The material illustrated in Figure 6 was obtained during field work at the Smithsonian Tropical Research Institute, Republic of Panama in 1984, funded by the National Geographic Society (U.S.), Grant No. 2578-82. Gary Brown prepared the line drawings. Figures 3 and 4 are reproduced from the Journal of General Physiology by copyright permission of the Rockefeller University Press.

References

Aggeler, J. and Werb, Z. (1982). Initial events during phagocytosis by macrophages viewed from outside and inside the cell: membrane-particle interactions and clathrin. *J. Cell Biol.* **94**, 613–623.

Anderson, R. G. W., Orci, L., Brown, M. S., Garcia-Segura, L. M. and Goldstein, J. L. (1985). Ultrastructural analysis of crystalloid endoplasmic reticulum in UT-1 cells and its disappearance in response to cholesterol. *J. Cell Sci.* **63**, 1–20.

Authi, K. S. and Crawford, N. (1985). Inositol 1,4,5-trisphosphate-induced release of sequestered Ca^{2+} from highly purified human platelet intracellular membranes. *Biochem. J.* **230**, 247–253.

Barlow, R. B. (1983). Circadian rhythms in the *Limulus* visual system. *J. Neurosci.* **31**, 856–870.

Barlow, R. B. and Chamberlain, S. C. (1980). Light and a circadian clock modulate structure and function in *Limulus* photoreceptors. *In* "The Effects of Constant Light on Visual Processes" (Ed. T. P. Williams and B. N. Baker), pp. 247–269. Plenum Press, New York.

Barlow, R. B., Bolanowski, S. J. and Brachman, M. L. (1977). Efferent optic nerve fibres mediate circadian rhythms in the *Limulus* eye. *Science* **197**, 86–89.

Barlow, R. B., Chamberlain, S. C. and Lerison, S. Z. (1980). *Limulus* brain modulates the structure and function of the lateral eyes. *Science* **201**, 1037–1039.

Battelle, B. A., Evans, J. A. and Chamberlain, S. C. (1982). Efferent fibres to *Limulus* eyes synthesise and release octopamine. *Science* **216**, 1250–1252.

Behrens, M. (1974). Photomechanical changes in the ommatidia of the *Limulus* lateral eye during light and dark adaptation. *J. comp. Physiol.* **89**, 45–57.

Behrens, M. and Krebs, W. (1976). The effect of light-dark adaptation on the ultrastructure of lateral eye retinula cells. *J. comp. Physiol.* **107**, 77–96.

Benshalom, C. and Reese, T. S. (1985). Ultrastructural observations on the cytoskeletal architecture of axons processed by rapid freezing and freeze-substitution. *J. Neurocytol.* **14**, 943–960.

Berridge, M. J. (1984). Inositol trisphosphate and diacylglycerol as second messengers. *Biochem. J.* **220**, 345–360.

Blest, A. D. (1978). The rapid synthesis and destruction of photoreceptor membrane by a dinopid spider: a daily cycle. *Proc. R. Soc. Lond.* B **200**, 463–483.

Blest, A. D. (1980). Photoreceptor membrane turnover in arthropods: comparative studies of breakdown processes and their implications. *In* "The Effects of Light on Visual Processes" (Ed. T. P. Williams and B. N. Baker), pp. 217–246. Plenum Press, New York.

Blest, A. D. (1985). The fine structure of spider photoreceptors in relation to function. *In* "The Neurobiology of Arachnids" (Ed. F. G. Barth), pp. 79–102. Springer-Verlag, Berlin-Heidelberg-New York-Tokyo.

Blest, A. D. and Day, W. A. (1978). The rhabdomere organization of some nocturnal pisaurid spiders in light and darkness. *Phil. Trans R. Soc. Lond.* B **283**, 1–23.

Blest, A. D. and Eddey, W. (1984). The extrarhabdomeral cytoskeleton in photoreceptors of Diptera. II. Plasmalemmal undercoats. *Proc. R. Soc. Lond.* B **220**, 353–359.

Blest, A. D. and Land, M. F. (1977). The physiological optics of *Dinopis subrufus* L. Koch: A fish-lens in a spider. *Proc. R. Soc. Lond.* B **196**, 197–222.

Blest, A. D. and Maples, J. (1979). Exocytotic shedding and glial uptake of photoreceptor membrane by a salticid spider. *Proc. R. Soc. Lond.* B **204**, 105–112.

Blest, A. D. and Price, D. G. (1981). A new mechanism for transitory, local endocytosis in photoreceptors of a spider, *Dinopis. Cell Tissue Res.* **217**, 267–282.

Blest, A. D. and Sigmund, C. (1985). The cytoskeletal architecture of interdigitated microvilli in the photoreceptors of some nocturnal spiders. *Protoplasma* **125**, 153–161.

Blest, A. D., Kao, L. and Powell, K. (1978a). Photoreceptor membrane breakdown in the spider *Dinopis*: the fate of rhabdomere products. *Cell Tissue Res.* **195**, 277–297.

Blest, A. D., Powell, K. and Kao, L. (1978b). Photoreceptor membrane breakdown in the spider *Dinopis*: GERL differentiation in the receptors. *Cell Tissue Res.* **195**, 277–297.

Blest, A. D., Price, G. D. and Maples, J. (1979). Photoreceptor membrane breakdown in the spider *Dinopis*: localization of acid phosphatases. *Cell Tissue Res.* **199**, 455–472.

Blest, A. D., Stowe, S. and Price, D. G. (1980a). The sources of acid hydrolases for photoreceptor membrane degradation in a grapsid crab. *Cell Tissue Res.* **205**, 229–244.

Blest, A. D., Williams, D. S. and Kao, L. (1980b). The posterior median eyes of the dinopid spider *Menneus. Cell Tissue Res.* **211**, 391–403.

Blest, A. D., Stowe, S. and Eddey, W. (1982a). A labile, Ca^{2+}-dependent cytoskeleton in the rhabdomeral microvilli of blowflies. *Cell Tissue Res.* **223**, 553–573.

Blest, A. D., Stowe, S., Eddey, W. and Williams, D. S. (1982b). The local deletion of a microvillar cytoskeleton from photoreceptors of tipulid flies during membrane turnover. *Proc. R. Soc. Lond.* B **215**, 469–479.

Blest, A. D., de Couet, H. G., Howard, J., Wilcox, M. and Sigmund, C. (1984a). The extrarhabdomeral cytoskeleton in photoreceptors of Diptera. I. Labile components in the cytoplasm. *Proc. R. Soc. Lond.* B **220**, 339–352.

Blest, A. D., Stowe, S. and de Couet, H. G. (1984b). Membrane turnover in the photoreceptors of arthropods. *Sci. Progr. Oxford* **69**, 83–100.

Bloom, J. W. and Atwood, J. L. (1981). Reversible ultrastructural changes in the rhabdom of the locust eye are induced by long-term light deprivation. *J. comp. Physiol.* **144**, 357–365.

Bok, D. (1985). Friedenwald Lecture. Retinal photoreceptor–pigment epithelium interactions. *Invest. Ophth. vis. Sci.* **26**, 1659–1694.

Boschek, C. B. and Hamdorf, K. (1976). Rhodopsin particles in the photoreceptor membrane of an insect. *Z. Naturforsch.* **31**, 763.

Brammer, J. D. and Clarin, B. (1976). Changes in volume of the rhabdom in the compound eye of *Aedes aegypti* L. *J. exp. Zool.* **195**, 33–40.

Brammer, J. D. and White, R. H. (1969). Vitamin A deficiency: effect on mosquito eye ultrastructure. *Science* **163**, 821–823.

Brammer, J. D., Stein, P. J. and Anderson, R. A. (1978). Effect of light and dark adaptation upon the rhabdom in the compound eye of the mosquito. *J. exp. Zool.* **206**, 151–156.

Brands, R., Snider, M. D., Hino, Y., Park, S. S., Gelboin, H. V. and Rothman, J. E. (1985). Retention of membrane proteins by the endoplasmic reticulum. *J. Cell Biol.* **101**, 1724–1732.

Brandenburger, J. L. and Eakin, R. M. (1985). Cytochemical localization of acid phosphatase in light- and dark-adapted eyes of a polychaete worm, *Nereis limnicola. Cell Tissue Res.* **224**, 623–628.

Bretscher, M. S. (1976). Direct lipid flow in cell membranes. *Nature, Lond.* **280**, 21–23.

Bruno, M. S., Barnes, S. N. and Goldsmith, T. S. (1977). The visual pigment and visual cycle of the lobster *Homarus. J. comp. Physiol.* **120**, 123–142.

Brown, J. E., Rubin, L. J., Ghalayini, A. J., Tarver, A. P., Irvine, R. F., Berridge, M. J. and Anderson, R. E. (1984). Myo-inositol polyphosphate may be a messenger for visual excitation in *Limulus* photoreceptor. *Nature, Lond.* **311**, 160–163.

Chamberlain, S. C. and Barlow, R. B. (1984). Transient membrane shedding in *Limulus* photoreceptors: control mechanisms under natural lighting. *J. Neurosci.* **4**, 2792–2870.

Chin, D. J., Luskey, K. L., Anderson, R. G. W., Faust, J. R., Goldstein, J. L. and Brown, M. S. (1982). Appearance of crystalloid endoplasmic reticulum in compactin-resistant Chinese hamster cells with a 500-fold elevation in 3-hydroxy-3-methylglutaryl Coenzyme A reductase. *Proc. nat. Acad. Sci. USA* **79**, 1185–1189.

Cowman, A. F., Zuker, C. S. and Rubin, G. M. (1986). An opsin gene expressed in only one photoreceptor cell type of the *Drosophila* eye. *Cell* **44**, 705–710.

Cronin, T. W. and Goldsmith, T. H. (1982). Quantum efficiency of the rhodopsin metarhodopsin conversion in crayfish photoreceptors. *Photochem. Photobiol.* **36**, 447–454.

Cronin, T. W. and Goldsmith, T. H. (1984). Dark regeneration of rhodopsin in crayfish photoreceptors. *J. Gen. Physiol.* **84**, 63–81.

Crouch, R. K. (1980). *In vitro* effects of light on the regeneration of rhodopsin. *In* "The Effects of Constant Light on Visual Processes" (Ed. T. P. Williams and B. N. Baker), pp. 309–318. Plenum Press, New York.

Cummins, D. R. and Goldsmith, T. H. (1986). Responses of crayfish photoreceptor cells following intense light adaptation. *J. comp. Physiol.* **158**, 35–42.

de Couet, H. G. (1984). Complex glycoproteins associated with the detergent-resistant membrane matrix of the rhabdomeral microvilli of crayfish photoreceptors. *Exp. Eye Res.* **39**, 279–298.

de Couet, H. G. and Blest, A. D. (1982). The retinal acid phosphatase of a crab, *Leptograpsus*: characterisation and relation to the cyclical turnover of photoreceptor membrane. *J. comp. Physiol.* **149**, 353–368.

de Couet, H. G. and Sigmund, C. (1985). Monoclonal antibodies to crayfish rhodopsin. I. Biochemical characterization and cross-reactivity. *Eur. J. Cell Biol.* **33**, 106–112.

de Couet, H. G. and Tanimura, T. (1987). Monoclonal antibodies provide evidence that rhodopsin in the outer rhabdomeres of *Drosophila melanogaster* is not glycosylated. *Eur. J. Cell Biol.* **44**, 50–56.

de Couet, H. G., Jablonski, P. P. and Perkin, J. L. (1986). Calmodulin associated with rhabdomeral microvilli of arthropods and squid. *Cell Tissue Res.* **244**, 315–319.

de Couet, H. G., Stowe, S. and Blest, A. D. (1984). Membrane-associated actin in the rhabdomeral microvilli of crayfish photoreceptors. *J. Cell Biol.* **98**, 834–846.

Eakin, R. M. and Brandenburger, J. L. (1985). Effects of light and dark on photoreceptors in the polychaete annelid *Nereis limnicola. Cell Tissue Res.* **242**, 613–622.

Eguchi, E. and Waterman, T. H. (1967). Changes in retinal fine structure induced in the crab *Libinia* by light and dark adaptation. *Z. Zellforsch.* **79**, 209–222.

Eguchi, E. and Waterman, T. H. (1976). Freeze-etch and histochemical evidence for cycling in crayfish photoreceptor membrane. *Cell Tissue Res.* **169**, 419–434.

Eguchi, E. and Waterman T. H. (1979). Longterm dark-induced fine structural changes in crayfish photoreceptor membrane. *J. comp. Physiol.* **131**, 191–203.

Europe-Finner, G. N. and Newell, P. C. (1986). Inositol 1,4,5-trisphosphate and calcium stimulate actin polymerisation in *Dictyostelium. J. Cell Sci.* **82**, 41–52.

Fein, A., Payne, R., Corson, W. D., Berridge, M. J. and Irvine, R. F. (1984). Photoreceptor excitation and adaptation by inositol 1,4,5-trisphosphate. *Nature, Lond.* **311**, 157–160.

Ferrell, J. E. and Huestis, W. H. (1984). Phosphoinositide metabolism and the morphology of human erythrocytes. *J. Cell Biol.* **98**, 1992–1998.

Ferrell, J. E., Lee, K. J. and Huestis, W. H. (1985). Membrane bilayer imbalance and erythrocyte shape: a quantitative assessment. *Biochemistry* **241**, 2849–2857.

Fliesler, S. J., Rapp, L. M. and Hollyfield, J. G. (1985). Photospecific degeneration caused by Tunicamycin. *Nature, Lond.* **311**, 575–577.

Geuze, H. J., Slot, J. W., Strous, G. J. A. M., Lodish, H. F. and Schwartz, A. L. (1983). Intracellular site of asialoglycoprotein receptor-ligand uncoupling: double-label immunoelectron microscopy during receptor-mediated endocytosis. *Cell* **32**, 277–287.

Gibbs, E. M. and Lienhard, G. E. (1984). Fluid-phase endocytosis by isolated rat adipocytes. *J. cell. Physiol.* **121**, 569–575.

Goldman, B. M. and Blobel, G. (1981). *In vitro* biosynthesis, Con A glycosylation and membrane integration of opsin. *J. Cell Biol.* **90**, 236–242.

Goldsmith, T. H. and Wehner, R. (1977). Restrictions on rotational and translational diffusion of pigment in the membranes of a rhabdomeric photoreceptor. *J. Gen. Physiol.* **70**, 453–490.

Goldstein, J. L., Anderson, R. G. W. and Brown, M. S. (1979). Coated pits, coated vesicles, and receptor-mediated endocytosis. *Nature, Lond.* **279**, 679–685.

Hafner, G. S. (1984). The *in vitro* uptake and distribution of ³H-fucose and ³H-mannose in the crayfish retina: a light and electron microscope autoradiographic study. *J. exp. Zool.* **231**, 199–210.

Hafner, G. S. and Bok, D. (1977). The distribution of ³H-leucine labelled protein in the retinula cells of the crayfish retina. *J. comp. Neurol.* **174**, 397–416.

Hafner, G. S. and Tokarski, T. R. (1986). The incorporation of ³H-fucose and ³H-mannose into the photopigment of the crayfish *Procambarus clarkii. Cell Tissue Res.* **243**, 109–115.

Hafner, G. S., Hammond-Soltis, G. and Tokarski, T. (1980). Diurnal changes of lysosome-related bodies in crayfish photoreceptor cells. *Cell Tissue Res.* **206**, 319–322.

Hafner, G. S., Tokarski, T., Jones, C. and Martin, R. (1982). Rhabdom degeneration in white-eyed and wild-type crayfish after long-term dark adaptation. *J. comp. Physiol.* **148**, 419–429.

Hamdorf, K. (1979). The physiology of invertebrate visual pigments. *In* "Vision in Invertebrates", Handbook of Sensory Physiology. (Ed. H. Autrum), pp. 145–224. Springer-Verlag, Berlin.

Hamdorf, K. and Kirschfeld, K. (1980). "Prebumps": evidence for double-hits at functional sub-units in a rhabdomeric photoreceptor. *Z. Naturforsch.* **35**, 197.

Harary, H. and Brown, J. E. (1984). Spatially non-uniform changes in intracellular calcium ion concentration. *Science* **224**, 292–294.

Hays, D. and Goldsmith, T. H. (1969). Microspectrophotometry of the visual pigment of the spider crab, *Libinia emarginata. Z. vergl. Physiol.* **65**, 218–232.

Henselman, R. A. and Cusanovich, M. A. (1974). The characterization of sodium cholate solubilised rhodopsin. *Biochemistry* **13**, 5199–5203.

Horridge, G. A., Duniec, J. and Marcelja, L. (1981). A 24-hour cycle in single locust and mantis photoreceptors. *J. exp. Biol.* **91**, 307–322.

Hubbard, A. L. and Stukenbrok, H. (1979). An electron microscope autoradiographic study of the carbohydrate recognition system in rat liver. II. Intracellular fates of the I-ligands. *J. Cell Biol.* **83**, 65–81.

Inoue, T. (1981). Pinocytotic vesicles observed by scanning electron microscopy. *Biomed. Res.* **2**, (Suppl.), 83–85.

Itaya, S. K. (1976). Rhabdom changes in the shrimp, *Palaemonetes. Cell Tissue Res.* **166**, 256–273.

Ivy, G. O., Schottler, F., Wenzel, J., Baudry, M. and Lynch, G. (1984). Inhibitors of lysosomal enzymes: accumulation of lipofuscin-like dense bodies in the brain. *Science* **226**, 985–987.

Kaplan, M. R. and Simoni, R. D. (1985a). Intracellular transport of phosphatidylcholine to the plasma membrane. *J. Cell Biol.* **101**, 441–445.

Kaplan, M. R. and Simoni, R. D. (1985b). Transport of cholesterol from the endoplasmic reticulum to the plasma membrane. *J. Cell Biol.* **101**, 446–453.

Kataoka, S. and Yamamoto, T. Y. (1985). Acyltransferase and acid hydrolase activities of the abalone photoreceptor cell. *Cell Tissue Res.* **241**, 59–65.

Kirschfeld, K. (1969). Absorption properties of photopigments in single rods, cones and rhabdomeres. *In* "Processing of Optical Data by Organisms and Machines" (Ed. M. Reichardt), pp. 166–173. Academic Press, New York.

Kirschfeld, K. (1982). Carotenoid pigments: their possible role in protecting against photo-oxidation in eyes and photoreceptor cells. *Proc. R. Soc. Lond.* B **216**, 71–85.

Kirschfeld, K. (1986). Activation of visual pigment. *In* "The Molecular Mechanism of Photoreception" (Ed. H. Stieve), pp. 31–49. Dahlem Konferenzen. Springer-Verlag, Berlin.

Kirschfeld, K. and Franceschini, N. (1969). Ein Mechanismus zur steurung des Lichtflusses in den Rhabdomeren des Komplexauges von *Musca. Kybernetik* **6**, 13–22.

Kirschfeld, K. and Vogt, K. (1980). Calcium ions and pigment migration in fly photoreceptors. *Naturwiss.* **67**, 516.

Krauhs, J. M., Mahler, H. R. and Moore, W. J. (1978). Protein turnover in photoreceptors of isolated *Limulus* eyes. *J. Neurochem.* **30**, 625–632.

Krinski, N. I. (1979). Carotenoid protection against oxidation. *Pure appl. Chem.* **51**, 649–660.

Land, M. F. (1980). Optics and vision in invertebrates. *In* "Handbook of Sensory Physiology" (Ed. H. Autrum), pp. 471–492. Springer-Verlag, Berlin.

Lane, N. J. and Swales, L. S. (1976). Interrelationships between Golgi, GERL, and synaptic vesicles in the nerve cells of insect and gastropod ganglia. *J. Cell Sci.* **22**, 435–453.

Laughlin, S. B., Blest, A. D. and Stowe, S. (1980). The sensitivity of receptors in the posterior median eye of the nocturnal spider *Dinopis. J. comp. Physiol.* **141**, 53–65.

Lo, M. V. C., Wong, F. and Pak, W. L. (1980). Increase in intracellular free calcium concentration of *Limulus* photoreceptors caused by metabolic inhibitor. *Vision Res.* **20**, 539–544.

Loew, E. R. (1976). Light and photoreceptor degeneration in the Norway lobster, *Nephrops norvegicus* (L.). *Proc. R. Soc. Lond.* B **193**, 31–44.

Loew, E. R. (1980). Visual pigment regeneration rate and susceptibility to photic damage. *In* "The Effects of Constant Light on Visual Processes" (Eds T. P. Williams and B. N. Baker), pp, 297–308. Plenum Press, New York.

Martin, R. L. and Hafner, G. S. (1986). Factors influencing the degradation of photoreceptor membrane in the crayfish. *Procambarus clarkii. Cell Tissue Res.* **243**, 205–212.

Meinecke, C. C. and Langer, H. (1984). Localization of visual pigments within rhabdoms of the compound eye of *Spodoptera exempta* (Insecta, Noctuidae). *Cell Tissue Res.* **238**, 359–368.

Meyer-Rochow, V. B. (1982). The divided eye of the isopod *Glyptonotus antarcticus*: effects of unilateral dark adaptation and temperature elevation. *Proc. R. Soc. Lond.* B **215**, 433–450.

Meyer-Rochow, V. B. and Eguchi, E. (1984). The effects of temperature and light on particles associated with crayfish visual membranes: a freeze-fracture analysis and electrophysiological study. *J. Neurocytol.* **13**, 935–959.

Meyer-Rochow, V. B. and Tiang, K. M. (1979). The effects of light and temperature on the structural organization of the eye of the Antarctic amphipod *Orchoneme plebs* (Crustacea). *Proc. R. Soc. Lond.* B **206**, 353–368.

Meyer-Rochow, V. B. and Tiang, K. B. (1982). Comparison between temperature-induced changes and effects caused by light–dark adaptation in the eyes of two species of Antarctic crustacean. *Cell Tissue Res.* **221**, 625–632.

Meyer-Rochow, V. B. and Waldvogel, H. (1979). Visual behaviour and the structure of dark and light-adapted larval and adult eyes of the New Zealand glow-worm *Arachnocampa luminosa* (Mycetophilidae: Diptera). *J. Insect Physiol.* **25**, 601–613.

Miller, W. H. (1975). Mechanisms of photoreceptor movement. *In* "Photoreceptor Optics" (Eds A. W. Snyder and R. W. Menzel), pp. 415–428. Springer-Verlag, Berlin.

Miller, W. H. and Cawthon, D. F. (1974). Pigment granule movement in *Limulus* photoreceptors. *Invest. Ophth. vis. Sci.* **13**, 401–405.

Mooseker, M. S. (1983). Actin binding proteins of the brush border. *Cell* **35**, 11–13.

Nässel, D. R. and Waterman, T. H. (1979). Massive, diurnally modulated photoreceptor membrane turnover in crab light and dark adaptation. *J. comp. Physiol.* **131**, 205–216.

Nathans, J., Thomas, D. and Hogness, D. S. (1986). Molecular genetics of human color vision: the genes encoding blue, green, and red pigments. *Science* **232**, 193–202.

Nilsson, H. L. and Lindstrom, M. (1983). Retinal damage and sensitivity loss of a light-sensitive crustacean compound eye (*Cirolana borealis*). *J. exp. Biol.* **107**, 277–292.

Nishizuka, Y. (1984). Turnover of inositol phospholipids and signal transduction. *Science* **225**, 1365–1370.

O'Brien, P. J. (1976). Rhodopsin as a glycoprotein: a possible role for the oligosaccharide in phagocytosis. *Exp. Eye Res.* **23**, 127–137.

O'Halloran, T., Beckerle, M. C. and Burridge, K. (1985). Identification of talin as a major cytoplasmic protein implicated in platelet activation. *Nature, Lond.* **317**, 449–451.

O'Tousa, J. E., Baehr, W., Martin, R. L., Hirsch, J., Pak, W. L. and Applebury, M. (1985). The *Drosophila ninaE* gene encodes an opsin. *Cell* **40**, 839–850.

Pallen, C. J. and Wang, J. H. (1985). A multifunctional calmodulin-stimulated phosphatase. *Arch. Biochem. Biophys.* **237**, 281–291.

Papermaster, D. S. and Schneider, B. G. (1982). Biosynthesis and morphogenesis of outer segment membrane in vertebrate photoreceptor cells. In "Cell Biology of the Eye" (Ed. D. S. McDevitt), pp. 475–531. Academic Press, New York.

Paulsen, R. and Schwemer, J. (1983). Biogenesis of blowfly photoreceptor membrane is regulated by 11-*cis*-retinal. *Eur. J. Biochem.* **137**, 609–614.

Pepe, I. M. and Baumann, F. (1972). Incorporation of ^{3}H-labelled leucine into the protein fraction in the honey-bee drone. *J. Neurochem.* **19**, 507–519.

Perrelet, A. (1972). Protein synthesis in the visual cells of the honeybee drone as studied with electron microscope autoradiography. *J. Cell Biol.* **55**, 595–605.

Piekos, W. B. and Waterman, T. H. (1983). Nocturnal rhabdom cycling and retinal haemocyte functions in crayfish (*Procambarus*) compound eyes. I. Light microscopy. *J. exp. Zool.* **225**, 209–217.

Pirenne, M. H. (1967) In "Vision and the Eye" (2nd edn). Chapman and Hall, London.

Rothman, J. E. (1981). The Golgi apparatus: two organelles in tandem. *Science* **213**, 1212–1219.

Ruggiero, M. and Lapetina, E. G. (1985). Leupeptin selectively inhibits human platelet responses induced by thrombin and trypsin: a role for proteolytic activation of phospholipase C. *Biochem. Biophys. Res. Comm.* **131**, 1198–1205.

Saibil, H. R. (1982). An ordered membrane–cystoskeleton network in squid photoreceptor microvilli. *J. molec. Biol.* **158**, 435–456.

Sato, S., Kato, M. and Toriumi, M. (1957). Structural changes of the compound eye of *Culex pipiens* var. *pallum* Coquillet in the process to dark adaptation. *Sci. Rep. Tohoku Univ.* **23**, 91–101.

Schinz, R. H., Lo, M. V. C., Larrivee, D. C. and Pak, W. L. (1982). Freeze-fracture study of the *Drosophila* photoreceptor membrane: mutations affecting membrane particle density. *J. Cell Biol.* **93**, 961–969.

Schnapp, B. J., Vale, R. D., Sheetz, M. P. and Reese, T. S. (1985). Single microtubules from squid axoplasm support bidirectional movement of organelles. *Cell* **40**, 455–462.

Schwemer, J. (1984). Renewal of visual pigment in photoreceptors of the blowfly. *J. comp. Physiol.* **154**, 535–547.

Schwemer, J. (1986). Turnover of photoreceptor membranes and visual pigment in invertebrates. In "The Molecular Mechanism of Photoreception" (Ed. H. Stieve). Dahlem Konferenzen, pp. 303–326. Springer-Verlag, Berlin.

Schwemer, J. and Henning, V. (1984). Morphological correlates of visual pigment turnover in photoreceptors of the fly, *Calliphora erythrocephala*. *Cell Tissue Res.* **236**, 293–303.

Sekar, M. C. and Hokin L. E. (1986). The role of phosphoinositides in signal transduction. *J. Membrane Biol.* **89**, 193–210.

Shelton, P. M. J., Gaten, E. and Chapman, C. J. (1985). Light and retinal damage in *Nephrops norvegicus* (L.) (Crustacea). *Proc. R. Soc. Lond.* B **226**, 217–236.

Skalska-Rakowska, J. M. and Baumgartner, B. (1985). Longitudinal continuity of the subrhabdomeric cisternae in the photoreceptors of the compound eye of the drone, *Apis mellifera*. *Experientia* **41**, 43–45.

Sluiman, H. J. (1984). A pathway of plasma membrane biogenesis bypassing the Golgi apparatus during cell division in the green alga *Cylindrocapsa geminella*. *J. Cell Sci.* **72**, 89–100.

Sly, W. S., Natowicz, M., Gonzalez-Nariega, A., Grubb, J. H. and Fisher, H. D. (1981). The role of the mannose-6-phosphate recognition marker and its receptor in the uptake and intracellular transport of lysosomal enzyme. *In* "Lysosomes and Lysosomal Storage Disease" (Eds J. W. Callahan and J. A. Lowden), pp. 131–145. Raven Press, New York.

Smith, P. J. S., Howes, E. A., Leech, C. A. and Treherne, J. E. (1986). Haemocyte involvement in the repair of the insect central nervous system after selective glial disruption. *Cell Tissue Res.* **243**, 367–374.

Snyder A. W. (1979). The physics of vision in compound eyes. *In* "Vision in Invertebrates", Handbook of Sensory Physiology (Ed. H. Autrum), pp. 225–313. Springer-Verlag, Berlin.

Stavenga, D. G. and Schwemer, J. (1984). Visual pigments of invertebrates. *In* "Photoreception and Vision in Invertebrates" (Ed. M. A. Ali), pp. 11–61. Plenum Press, New York.

Stowe, S. (1980). Rapid synthesis of photoreceptor membrane and assembly of new microvilli in a crab at dusk. *Cell Tissue Res.* **211**, 419–440.

Stowe, S. (1981). Effects of illumination changes on rhabdom synthesis in a crab. *J. comp. Physiol.* **142**, 19–25.

Stowe, S. (1982). Rhabdom synthesis in isolated eyestalks and retinae of the crab *Leptograpsus variegatus*. *J. comp. Physiol.* **148**, 313–321.

Stowe, S. (1983a). Light-induced and spontaneous breakdown of the rhabdoms in a crab at dawn: depolarisation *versus* calcium levels. *J. comp. Physiol.* **153**, 365–375.

Stowe, S. (1983b). Phagocytosis of photoreceptor membrane at dawn in a crab. *Cell Tissue Res.* **234**, 463–467.

Stowe, S., Fukudome, H. and Tanaka, K. (1986). Membrane turnover in crab photoreceptors studied by high-resolution scanning electron microscopy, and by a new technique of thick-section transmission electron microscopy. *Cell Tissue Res.* **245**, 51–60.

Strausfeld, N. J. (1976). *In* "Atlas of an Insect Brain". Springer-Verlag, Berlin.

Tilney, L. G. (1979). Actin, motility and membranes. *In* "Membrane Transduction Mechanisms" (Eds R. A. Cone and J. E. Dowling). Raven Press, New York.

Toh, Y. and Waterman, T. H. (1982). Diurnal changes in compound eye fine structure in the blue crab *Callinectes*. I. Differences between noon and midnight retinae on a LD 11:13 cycle. *J. Ultrastruct. Res.* **78**, 40–59.

Trowell, S. C. (1984). Cytochemical distribution and biochemistry of a novel phosphatase in the photoreceptive microvilli of a crab. *Eur. J. Cell Biol.* **36**, 277–285.

Trowell, S. C. (in press). Partial purification and characterization of the 4-nitrophenylphosphatase activity of invertebrate photoreceptive microvilli. Absence of *in vitro* rhodopsin phosphatase activity. *Comp. Biochem. Physiol.*

Tsuda, M., Tsuda, T., Terayama, Y., Fukada, Y., Akino, T., Yamanaka, G., Stryer, L., Katada, T., Ui, M. and Ebrey, T. (1986). Kinship of photoreceptor G-protein with vertebrate tranducin. *FEBS Letts.* **198**, 5–10.

Tsung, P. K. and Lombardini, J. B. (1985). Identification of low Ca^{2+}- and high Ca^{2+}-requiring neutral proteases in rat retina. *Exp. Eye Res.* **41**, 97–103.

Tuurala, O. and Lehtinen, A. (1971a). Über die Einwirkung von Licht und Dunkel auf die Feinstruktur der Lichtsinneszellen der Assel, *Oniscus asellus* L. 1. Länge der Microvilli und Anzahl der multivesicularen Körper. *Ann. Acad. Sci. Fenn. Ser. A IV Biol.* **176**, 1–9.

Tuurala, O. and Lehtinen, A. (1971b). Über die Einwirkung von Licht und Dunkel auf die Feinstruktur der Lichtsinneszellen der Assel, *Oniscus asellus* L. 2. Microvilli und multivesicularen Körper nach starker Belichtung. *Ann. Acad. Sci. Fenn. Ser. A IV Biol.* **177**, 1–8.

Tuurala, O. and Lehtinen, A. (1974). Inkorpierung des tritiummarkierten Leucins in der Sehzellen von *Oniscus asellus* L. (Isopoda, Oniscoidea). *Ann. Zool. Fenn.* **11**, 135–140.

Vale, R. D., Schnapp, B. J., Reese, T. S. and Sheetz, M. P. (1985). Movement of organelles along filaments dissociated from the axoplasm of the squid giant axon. *Cell* **40**, 449–454.

Varela, F. G. and Porter, K. R. (1969). Fine structure of the visual system of the honey-bee (*Apis mellifera*). I. The retina. *J. Ultrastruct. Res.* **29**, 236–259.

Vogt, K. (1983). Is the fly visual pigment a rhodopsin? *Z. Naturforsch.* **38**, 329–333.

Vogt, K. (1984). The chromophore of the visual pigment in some insect orders. *Z. Naturforsch.* **39**, 196–197.

Vogt, K. and Kirschfeld, K. (1984). Chemical identity of the chromophore of fly visual pigment. *Naturwiss.* **71**, 211–213.

Walter, M. G. and Biessmann, H. (1984). Intermediate-sized filaments in *Drosophila* tissue culture cells. *J. Cell Biol.* **99**, 1468–1477.

Walz, B. (1982a). Ca^{2+}-sequestering smooth endoplasmic reticulum in an invertebrate photoreceptor. I. Intracellular topography as revealed by OsFeCN staining and *in situ* Ca accumulation. *J. Cell Biol.* **93**, 839–848.

Walz, B. (1982b). Ca^{2+}-sequestering smooth endoplasmic reticulum in an invertebrate photoreceptor. II. Its properties as revealed by microphotometric measurements. *J. Cell Biol.* **93**, 849–859.

Walz, B. (1982c). Calcium-sequestering smooth endoplasmic reticulum in retinula cells of the blowfly. *J. Ultrastruct. Res.* **81**, 240–248.

Waterman, T. H. and Piekos, W. B. (1981). Light and time correlated migration of invasive haemocytes in the crayfish compound eye. *J. exp. Zool.* **217**, 1–14.

Waterman, T. H. and Piekos, W. B. (1983). Nocturnal rhabdom cycling and retinal haemocyte functions in crayfish (*Procamabrus*) compound eye. II. Transmission electron microscopy and acid phosphatase localisation. *J. exp. Zool.* **225**, 219–231.

White, R. H. (1967a). The effects of light and light-deprivation upon the ultrastructure of the larval mosquito eye. II. The rhabdom. *J. exp. Zool.* **166**, 405–425.

White, R. H. (1967b). The effects of light and light-deprivation upon the ultrastructure of the larval mosquito eye. III. Multivesicular bodies and protein uptake. *J. exp. Zool.* **169**, 261–278.

White, R. H. and Lord, E. (1975). Diminution and enlargement of mosquito rhabdom in light and darkness. *J. gen. Physiol.* **65**, 583–598.

White, R. H. and Michaud, N. A. (1981). Disruption of insect photoreceptor membrane by divalent ions: dissimilar sensitivity of light- and dark-adapted mosquito rhabdomeres. *Cell Tissue Res.* **216**, 403–412.

White, R. H., Gifford, D. and Michaud, N. A. (1980). Turnover of photoreceptor membrane in the larval mosquito ocellus: rhabdomeric coated vesicles and organelles of the vacuolar system. *In* "The Effects of Constant Light on Visual Processes" (Eds T. P. Williams and B. N. Baker), pp. 271–296. Plenum Press, New York.

Whittle, A. (1976). Reticular specialisations in photoreceptors: a review. *Zool. Scripta.* **5**, 191–206.

Wileman, T., Harding, C. and Stahl, P. (1985). Receptor-mediated endocytosis. *Biochem. J.* **232**, 1–14.

Williams, D. A. (1980a). Ca^{++}-induced structural changes in photoreceptor microvilli of Diptera. *Cell Tissue Res.* **206**, 225–232.

Williams, D. S. (1980b). Organisation of the compound eye of a tipulid fly during the day and night. *Zoomorphology* **95**, 85–104.

Williams, D. S. (1982a). Rhabdom size and photoreceptor membrane turnover in a muscoid fly. *Cell Tissue Res.* **226**, 629–639.

Williams, D. S. (1982b). Ommatidial structure in relation to turnover of photoreceptor membrane in the locust. *Cell Tissue Res.* **225**, 595–617.

Williams, D. S. (1982c). Photoreceptor membrane shedding and assembly can be initiated locally within an insect retina. *Science* **218**, 898–900.

Williams, D. S. (1983). Changes of photoreceptor performance associated with the daily turnover of photoreceptor membrane in locusts. *J. comp. Physiol.* **105**, 509–519.

Williams, D. S. and Blest, A. D. (1980). Extracellular shedding of photoreceptor membrane in the open rhabdom of a tipulid fly. *Cell Tissue Res.* **205**, 423–438.

Wilson, M. (1975). Angular sensitivity of light and dark-adapted locust retinula cells. *J. comp. Physiol.* **97**, 323–328.

Winterhager, E. and Stieve, H. (1982). Effect of hyper- and hyposmotic solutions on the structure of the *Astacus* retina. *Cell Tissue Res.* **223**, 267–280.

Wu, E-S., Tank, D. W. and Webb, W. W. (1982). Unconstrained lateral diffusion of concanavalin A receptors on bulbous lymphocytes. *Proc. Nat. Acad. Sci.* **79**, 4962–4966.

Young, R. W. (1967). The renewal of photoreceptor cell outer segments. *J. Cell Biol.* **33**, 61–72.

Young, R. W. (1976). Visual cells and the concept of renewal. *Invest. Ophthal. vis. Sci.* **15**, 700–725.

Zhu, H. and Kirschfeld, K. (1984). Protection against photodestruction in fly photoreceptors by carotenoid pigments. *J. comp. Physiol.* **154**, 153–156.

Zucker, C. S., Cowman, A. F. and Rubin, G. M. (1985). Isolation and structure of a rhodopsin gene from *D. melanogaster. Cell* **40**, 851–858.

Addendum

Since this account was completed, several molecular genetic analyses of *Drosophila* eye mutants have been published, mostly as abstracts from conference proceedings. Three are relevant to an understanding of rhabdomeral turnover. They modify some of the subjects discussed in the preceding Review.

(1) It was noted earlier that *ninaC* mutants exhibit deformed rhabdomeral microvilli and an abnormality of the microvillar cytoskeleton (Section 8.1). Two eye-specific polypeptides of 130 and 170 KD encoded by the gene may contribute to regulation of the cytoskeleton, or may be a component of it (Matsumoto *et al.*, 1986). Montell *et al.* (1986) have since examined the *ninaC* proteins, and found them to be of 130 and 155 KD. Each of the two proteins incorporates two distinctive domains. One (from amino acid sequence homologies) resembles various protein kinases, the other resembles the heavy myosin chain.

(2) O'Tousa *et al.* (1986) have examined $ora^{JK\ 84}$, in which rhabdomeres of R_{1-6} degenerate rapidly after eclosion. $Ora^{JK\ 84}$ proved to contain two mutations: one (*ort*) possesses normal photoreceptor structures in the homozygote, but lacks the transient components of the electroretinogram. The other lies within the *ninaE* gene which encodes the R_{1-6} rhodopsin. It is responsible for the early degeneration of R_{1-6} rhabdomeres. The conclusion that the presence of rhodopsin plays a vital role in maintaining the microvillar architecture of rhabdomeres appears to be at variance with the findings of Schinz *et al.* (1982), summarized in Section 5.2.

Each of these studies illustrates the potential power of molecular genetic techniques for the analysis of microvillar stability and the mechanisms that regulate phototransductive membrane turnover. In addition, each hints at the complexity of the molecular pathways that still confront us. I am indebted to Dr G. L. G. Miklos for drawing my attention to these abstracts, and for discussing them with me.

Additional references

Matsumoto, H., Ozaki, K., Isono, K., Randall, L., Floreani, M., Larrivee, D., Pye, Q. and Pak, W. L. (1986). *nina*-Class of mutations affecting photoreceptor-specific gene functions. *In*: Abstracts of *Molecular Neurobiology of Drosophila*, Cold Spring Harbor Laboratory, Cold Spring Harbor, New York. p. 17.

Montell, C., Zuker, C. and Rubin, G. (1986). Molecular analysis of phototransduction: identification and characterization of the genes encoding *TRP*, *NINAC* and two R_7-specific opsins. *In*: Abstracts of *Molecular Neurobiology of Drosophila*. Cold Spring Harbor Laboratory, Cold Spring Harbor, New York. p. 17.

O'Tousa, J. E., Leonard, D. S. and Pak, W. L. (1986) Morphological defects in $ora^{JK\ 84}$ photoreceptors caused by mutation in R_{1-6} opsin gene of *Drosophila*. Cold Spring Harbor Laboratory, Cold Spring Harbor, New York. p. 71.

Honey Bee Learning

James L. Gould* and William F. Towne†

*Department of Biology, Princeton University, Princeton, NJ 08544, USA
†Department of Biology, Kutztown University, Kutztown, PA 19530, USA

1   Aims and scope

The aim of this review is to integrate behaviouristic terminology and insights with the ethological perspective on learning. Because so much is known about its natural history, and because it is such a convenient experimental "wild" animal, the honey bee is a natural choice to illustrate the value of this convergence. In addition, there has been a virtual explosion of work on honey bee learning and memory in the past 15 years, and the insights these studies have provided are probably broadly applicable to insects in general.

2   Ethology of learning

Traditionally, learning has been studied either in the field by ethologists under natural or semi-natural conditions, or in the laboratory by behaviour-

ADVANCES IN INSECT PHYSIOLOGY VOL. 20
ISBN 0–12–024220–6

istic psychologists under controlled but relatively unnatural conditions. Among the ethologists, the emphasis has been on the adaptiveness of learning and its interactions with instinctive behaviour; behaviourists, on the other hand, have concentrated more on discovering general rules of learning.

In their review of natural learning, Gould and Marler (1984) made the point that learning, like all other aspects of behaviour, appears to have evolved to solve the specific problems the members of a species face in making a living in their particular niche. Their species-specific learning is generally known as selective learning. Where innate recognition and reaction are possible, they have evolved; where learning is essential, it has evolved, Selective learning is best suited for dealing with what they call the "predictably unpredictable": general situations in which learning is going to be needed and where the nature of the object or individual to be learned is predictable, but the individual details of the object or individual are either too complex, too numerous, or too unpredictable to be prewired. An example that illustrates this principle involves enemy learning by birds.

Birds face two sorts of predators: large hunters such as hawks that can take adults, and nest predators such as crows that prey on eggs or nestlings. Birds are equipped with an innate ability to spot and react to a pair of forward-facing eyes, a typical feature of predators (Arduino and Gould, 1984), but except for owls few nest predators are thus equipped. Instead of needing to learn from direct experience by observing its own eggs being destroyed, or having an innate "field guide" to potential predators, or being bound by a straightforward fear of everything animate, birds learn which animals are dangerous and which are benign. Curio and his colleagues (Curio and Vieth, 1978) have shown that this learning system is centered on the mobbing call, a signal which is innately produced and innately recognized. When a bird hears the call, it observes the creature being mobbed and commits its appearance to memory. When it sees another member of that species in the future, it calls and mobs. Curio cleverly tricked a blackbird into mobbing a stuffed nectar-feeding bird, and it passed the aversion to other birds of several species. Curio was even able to teach a "hatred" of detergent bottles to a laboratory population.

The essential features of the enemy-learning program in song birds are as follows: (1) the situation is predictable (nesting birds are subject to nest predation) and the context of learning can be innately specified (it occurs in the presence of the mobbing call); (2) the things to be learned (the nest predators) are either too numerous, too unpredictable (from one region to another), or too visually complex to be innately stored. In this case, they are all three. This innate direction of learning is the general theme of selective learning, and provides an organizing principle for our discussion of honey bee learning.

3   Psychology of learning: classical conditioning

The most familiar type of associative learning, classical or Pavlovian conditioning, involves the formation of an association between two stimuli, an
unconditioned stimulus (US), which reflexively elicits an unconditioned response (UR), and a conditioned stimulus (CS), which initially does not. In a
typical training experiment, each presentation of the CS is followed quickly
by a presentation of the US (which elicits the UR); in time the CS alone
comes to elicit a response, the conditioned response (CR). The CR and UR
are often, though not always, identical, the important difference being that
the CR is produced by the initially ineffective CS. Pavlovian associations can
form whenever the CS is sufficiently predictive of the US, although this is by
no means always the case. The classic textbook example of Pavlovian conditioning was, of course, Pavlov's dog, which when repeatedly presented with a
sound (CS) followed by food (US) eventually salivated (CR) in response to
the sound alone (Pavlov, 1927).

Ethologists have been studying learning for almost as long as Pavlov and
his descendents—von Frisch (1914), for example, showed that bees can learn,
and therefore must be able to see, the colours of flowers, much as Pavlov's experiments showed that his dog could both learn and hear. It may therefore be
useful to restate classical conditioning in ethological terms: the innately
recognized US is a "sign stimulus", the UR "reflex" is a motor program, and
the learned CS is, loosely speaking, a search image. The adaptive value of this
prewired ability to recognize associations seems clear: a simple, innate sign
stimulus comes to be replaced with what is potentially a far more detailed
picture of the object bearing the sign stimulus. The colour, location and
shape of the best food items, for example, cannot often be predicted with precision by an animal which has just entered the world, but the animal's associative learning programs ensure that it will quickly come to recognize, from
a world full of colours, locations, and shapes, those that reliably predict the
presence of food.

Students of classical conditioning at one time assumed that if the stimuli
could be sensed, any association could be learned by any animal. This
natural extension of the "blank slate" view of animals was never taken ser-
iously by ethologists, and has since been abandoned by behaviourists as well.
It is now generally recognized that most associative learning occurs only in
specific behavioural contexts and with strong stimulus and response biases.
Pigeons, for example, more readily learn to associate sounds than visual cues
with danger (electric shock), yet learn to associate visual cues to the near
exclusion of auditory cues with food (Foree and LoLordo, 1973; LoLordo,
1979). It is as though pigeons are wired to "expect" that certain cues are more
likely to be useful in certain contexts and so should be attended to first. The

evolutionary logic of this seems obvious: an animal programmed with the right kind of learning biases is likely to learn better and faster, and so have an edge in the struggle for survival. An animal with no biases would clearly be at a disadvantage, assuming that the situation is of the sort we have called predictably unpredictable.

One variant of classical conditioning, alpha conditioning, even suggests how nervous systems might be designed to yield these learning biases (Carew *et al.*, 1984; Gould and Marler, 1984). Alpha conditioning is the enhancement, through classical conditioning procedures, of a pre-existing response to a CS. Pavlov's dog, for example, salivated slightly in response to certain novel stimuli, including sounds. Alpha conditioning, then, probably involves the mere strengthening of pre-existing neural connections. Indeed, classical conditioning of the gill-withdrawal response in the marine mollusc *Aplysia*, perhaps the best-studied case of associative learning at the cellular level, seems to work in just this way (Hawkins *et al.*, 1983). It is tempting to speculate that most or all classical conditioning is actually true or covert alpha conditioning (Carew *et al.*, 1984; Gould and Marler 1984) in which the neural circuitry supporting the CR must be innately prewired. This is by far the best explanation so far for the stimulus and response biases so common in associative learning.

We do not wish to leave the impression that all associative learning increases the likelihood of an animal's responding to the CS. Just as the probability of responding to a CS increases when it reliably predicts the US, the probability of responding generally decreases when the CS reliably predicts the absence of the US. For example, if the CS and US are presented separately on alternate training trials, subsequent learning of a positive association between the stimuli is often impaired; this phenomenon is known as conditioned inhibition (Schwartz, 1984). Conditioned inhibition resulting from explicitly unpaired presentations of the CS and US is very familiar from laboratory work with vertebrates, and the same basic phenomenon has been observed in laboratory experiments with bees (Bitterman *et al.*, 1983).

Several other processes involving associative learning which, like conditioned inhibition, were first discovered in lab studies of vertebrates and later found in bees and other invertebrates, strongly support two important conclusions: (1) animals in general are genetically endowed with strong learning predispositions, as mentioned previously and (2) the basic phenomenology of associative learning is remarkably similar in vertebrates and invertebrates. Among the latter, this is particularly true for molluscs (Sahley, 1984; Sahley *et al.*, 1984) and bees (Menzel, 1983; Bitterman *et al.*, 1983; Menzel and Bitterman, 1983; Abramson, 1986). Both vertebrates and invertebrates have been reasonably well studied. We will discuss only two of these additional processes: overshadowing in compound conditioning experiments and the temporal phases of associative memory.

Compound conditioning is simple Pavlovian conditioning in which there are two CSs, for instance a sound and a light, presented simultaneously and, as usual, just before the US during training. Even if the two CSs support roughly equal conditioning when presented alone, when the two are presented as a compound, one often overshadows the other; one is learned while the other is not. Overshadowing is common among vertebrates (Mackintosh, 1974) and can probably be taken as another type of adaptive learning predispositon. In keeping with our second point above, overshadowing has been found in honey bees; although both odour and colour can be learned as markers of food, colour is overshadowed (though not completely) by odour (Couvillon and Bitterman, 1982).

The temporal phases of memory appear to be remarkably similar in virtually all animals, including bees (Menzel, 1983, 1984). An early phase, called short-term memory, typically lasts for minutes, although the actual time course can vary greatly. It is easily disrupted or erased by electroconvulsive shock or by any experience which conflicts with the information just learned. In bees, the stored association of one colour with food is erased by the presentation of a second colour with food only if the second colour is presented within about 5 minutes of the initial training, while the memory trace is in the short-term phase (Menzel, 1979). The second memory phase, called long-term memory, develops later, lasts much longer (hours to weeks), and once formed is resistant to interference from electroconvulsive shock and conflicting experience. The formation of long-term memory requires new protein synthesis in all vertebrate and invertebrate species so far tested, while the formation of short-term memory does not (Goelet et al., 1986).

In bees at least, there appears to be a rapid transfer from short-term to long-term memory when similar learning events follow one another in rapid sequence, as if the memory trace is passed on when the short-term memory is full of consistent information. This suggests that short-term memory may be part of an adaptive information-processing strategy rather than a physiologically necessary detour that the information must take on its way to long-term storage. That is, short-term memory is a way to temporarily hold information while waiting to see if it will be corroborated (Erber, 1975a, b; Menzel, 1979, 1984).

We will discuss the importance of learning predispositions further below, but what can be made of the numerous functional and mechanistic similarities between associative learning in vertebrates and invertebrates? In short, we do not know. It could well be, and this is the hope of those who study the cellular basis of learning in invertebrates, that the similarities reflect homologies at the cellular level (Menzel, 1983; Sahley et al., 1984). On the other hand, the ability of animals to extract predictive relationships from the bio-logically important features of their environments is clearly highly adaptive, and it would not be surprising if it had arisen more than once.

4 Psychology of learning: operant conditioning

Trial-and-error learning, called operant or instrumental conditioning by psy-chologists, is associative learning in which behaviours (motor programs), rather than conditioned stimuli, come to be associated with external events (USs, such as food, water, or electric shock in lab experiments). The nature and result of the learning in the lab can be summarized in part by Thorn-dike's (1898) famous "law of effect": behaviours which are followed closely in time by rewards are produced with higher frequency in the future, whereas those which are followed by punishment are produced less frequently. The classic example of trial-and-error learning involves Thorndike's cat, which eventually learned how to open a simple latch on its cage door to get a bit of food placed in view but out of reach through the cage bars.

In acquiring the learned behaviour, an animal typically begins by perform-ing a set of innate motor patterns appropriate to the situation given the ani-mal's natural history. A hungry cat will walk, paw and prod objects in its environment. If in the course of these activities the cat accidentally releases the latch on its cage door, escapes, and is therefore able to eat the food, the next time it encounters the same situation it may perform the same set of be-haviours mainly near the cage door and will therefore succeed a bit sooner. Eventually, it will learn to open the latch quickly and economically.

Since not all motor patterns in the repertoire of an animal occur in a given behavioural context, only a subset of the animal's innate motor patterns can be involved in the learned response in a given training situation (Breland and Breland, 1961; Staddon and Simmelhag, 1971). Moreover, of the subset of available motor patterns, some are far more easily associated with certain USs, and, further, the exact form that the final motor pattern will take is often inflexible. It is very easy, for example, to train pigeons to peck a key to obtain a food reward, but cooing, treadle-hopping, and wing-flapping cannot be easily conditioned when food is the US (Schwartz, 1984). Furthermore, if food is the reward (US), pigeons will peck the key with a "food peck" (rapid and with the bill open), whereas if the reward is water, the pigeons will give a "water peck" (slower and with the bill nearly closed). One cannot teach a pigeon to give a water peck for food or a food peck for water (Moore, 1973). This is contrary to Thorndike's law of effect and Skinner's (1938) influential proposal that the final form of the learned motor pattern is strictly dependent upon the experimental contingency. However, it is not hard to understand, given that the pigeons are wired to help them associate with food or water those responses which will best help the birds to obtain "rewards" under natural conditions.

Trial-and-error learning appears to be widespread and important in nature, and several studies suggest that the learning may have even more innate guidance than the above examples suggest. Most animals must learn how to perform a variety of complex motor tasks with speed and precision, and a general picture of how this is done is beginning to emerge. The copulation of male rhesus monkeys, the food-burying behaviour of squirrels, and the vocalizations of many birds are all assembled through trial-and-error learning; they all take practice. However, in each case, the animals do not need to start entirely from scratch when learning how to perform these tasks. They are prewired with the appropriate sign stimulus recognition circuits and the various motor subunits, and they "know" that the subunits belong together; they need only assemble them in the appropriate way (Gould and Marler, 1984). This type of learning seems to differ from operant learning in the laboratory only in that the "rewards" are internal and the resultant motor patterns more complex. Further, it resembles laboratory learning in that the motor patterns, once assembled, often become "hardwired". A bird sings perfectly, even if it is deafened, after its song has "crystallized" (Marler, 1984). A person does not need to think about tying his shoelaces once the motor pattern is "modularized" (Connolly, 1977). Rats running mazes will run into newly erected barriers or will stop once they have run the usual distance down an alley to a food dish, even though the dish has only been moved a short distance farther along (Schwartz, 1978). In each case, once the learning is complete the animal's attention is freed to attend to other tasks even as the motor pattern is being executed. This basic strategy for learning motor patterns seems likely to be general.

The study of stimulus–stimulus associations (classical conditioning) in invertebrates is, as Sahley (1984) put it, still in its infancy. The study of stimulus–response associations (operant or trial-and-error learning) in invertebrates, on the other hand, with the exception of one or two recent findings discussed below, has not yet begun at all. It would be surprising if the work turned up any fundamental differences between vertebrates and invertebrates, but this is an empirical question awaiting study.

5 The foraging cycle

Honey bees make their living by gathering nectar and pollen from flowers (for a review see Gould, 1982). Certain bees, called scouts, leave the hive on exploratory flights, searching for new sources of food; others, usually called foragers, make round trips to familiar sources of food. Yet other bees, known as recruits, leave after attending a dance in the hive which specifies the distance and direction of a food source discovered by a forager; the recruits usually also gain information on the odour and quality of the food from the dancer. The task faced by each of these three classes of bee is slightly different.

For the scout, the flight path will, of necessity, be circuitous. Nevertheless, when food is discovered, or when the bee "gives up", it must be able to fly directly back to the hive. Given the large distances involved and the poor visual resolution of bees, they cannot rely on being able to see the home hive. It appears that on the outward flight the scout keeps track of the direction and length of each leg of its search, integrating the legs either as it goes or at the end. This capacity requires a reference system, and von Frisch (1967) showed long ago that the primary cue is the sun's azimuth. However, to use the sun the bees must be able to recognize it and compensate for its varying apparent movement through the sky. Brines and Gould (1979) showed that the recognition of the sun is not learned, but instead is based on a sign stimulus (its low percentage of UV light). The direction of movement is learned (Lindauer, 1959); this is a necessity since bees live in both the northern and southern hemispheres. Moreover, bees also use learning to compensate for the sun's change in azimuth. Where prominent, distinguishable landmarks are available, they memorize the azimuth for each time of day (Dyer and Gould, 1981, 1983); where useful landmarks are not available, the bees use their memory of the most recent rate of azimuth movement and "extrapolate" to allow for the passage of time (Gould, 1980). This extrapolation involves taking a running average of azimuth movement over periods of about 40 minutes (Gould, 1984b). Bees are able to substitute patterns of polarized light in the sky when the sun is not visible, although whether learning is involved in this process is the subject of some contention (Dyer and Gould, 1983).

Foragers, on the other hand, are usually familiar with the route they are taking to the food and frequently come to depend on prominent landmarks along the route for navigation (von Frisch and Lindauer, 1954). This is not to say that they are oblivious to celestial cues; if the bees are moved to a location with similar landmarks but which lie in a different direction, they will navigate with reference to the landmarks but orient their dances with reference to the sun (Dyer and Gould, 1981).

Recruits, having obtained a set of radial coordinates specifying the location of food, are generally thought to use celestial information to guide themselves to the vicinity of the goal; once there, they use their memory of the dancer's odour to find the food source and track it upwind. It is also possible that recruits use landmarks to find food; in one experiment (Gould and

Gould, 1982) recruits attending dances specifying the middle of an adjacent lake refused to leave the hive, but were readily recruited by dances to the same food and at the same distance corresponding to a site at the edge of the lake. One interpretation of this result is that the recruits were able to "place" the coordinates on a mental map and judge the relative "plausibility" of the two sites. Experiments discussed in a later section demonstrate that bees do possess mental maps, although whether they can use them in this way is not yet known.

Once scouts and recruits are in the field searching for food, they do not land on every type of object they encounter, such as blades of grass, for example. Instead, they react innately to objects that have certain special characteristics (Gould, 1984a); these include targets that are colourful (especially violet objects) and contrast well with the background, targets that have floral odours (especially geraniol), targets that are visually "busy" (have a high spatial frequency, as do flowers with several petals), and so on. Targets with one or, better yet, more of these characteristics receive the most attention and so bees are led to land and explore plausible flower-like objects.

Bees are flower-constant, specializing in a particular species, at least at given times of day. To do this, they must be able to remember enough about a flower to be able to recognize it upon subsequent encounters. The nature of this flower memory is the subject of the rest of this review.

6 Odour learning

Von Frisch (1919) showed almost 70 years ago that free-flying bees quickly and accurately learn to associate a great variety of odours with food. Indeed, it now appears that no known odours, even ones that are initially repellent, will fail to become attractive to bees when presented along with food, although flower-like odours are learned considerably faster and better than others (Menzel, 1985). Geraniol, a component of the Nasanov gland pheromone used by bees to mark especially rich food sources, is learned the fastest of all, yielding over 95% correct choices in two-odour choice tests after a single training trial. Most floral odours are learned almost as well. The adaptive value of this rapid learning seems clear: the most distinctive and reliable marker of most flower species is their odour, so by quickly memorizing odours, bees can reliably relocate profitable blossoms, even among an array of different species which might be similar in colour.

Because odour learning is so quick and reliable, it has been very useful in studying both the olfactory abilities of bees (which turn out to be strikingly similar to our own; von Frisch, 1967) and the psychological phenomenology of bee learning (Couvillon and Bitterman, 1980, 1982, 1984; Couvillon et al.,

1983; Bitterman *et al.*, 1983). In addition, comparative studies of odour learning in three races of *Apis mellifera* and another species of honey bee, *Apis cerana*, have shown that these different races and species learn different floral odours best (Koltermann, 1973). The highest-ranking floral odours for each race or species seem to correspond to the floral odours present in the native geographical areas of the bees (Lindauer, 1976). Although odour learning in free-flying bees has provided insights into the natural history and psychology of bee learning, the physiological mechanisms of odour learning must be studied with restrained bees in the laboratory. Fortunately, there is a procedure, classical conditioning of the proboscis-extension reflex, which gives strong one-trial learning and has permitted physiological investigation of the brain during such learning. A brief (e.g. 3 second) puff of odour (CS) is followed by touching one of the bee's antennae with a drop of sugar water (US); this elicits extension of the proboscis (UR) and drinking. Typically, between 60% and 80% of the bees show conditioned responses after only a single training trial (Bitterman *et al.*, 1983; Menzel and Bitterman, 1983). Menzel, Erber, and colleagues (Erber, 1980, 1981, 1983, 1984; Erber *et al.*, 1980; Menzel, 1984; Menzel *et al.*, 1974) have used this procedure to trace the anatomical location of the short-term memory over time and to begin the search for neural correlates of learning. By cooling different parts of the brain at varying intervals after the odour–food pairing, the neural activity associated with short-term memory has been shown to begin in the antennal lobes (the sensory neuropil of the antennae), and to proceed about 2 minutes later to the alpha lobes of the mushroom bodies, and another 2 minutes later to the calyces. After this, chilling has no effect anywhere. Furthermore, by training only one antenna, Menzel and colleagues showed that the process is initially unilateral, but that the information leaving the alpha lobe is sent to both calyces. Finally, and perhaps of most interest, is that Erber (1981) found neurons in the median protocerebrum that showed enhanced olfactory responsiveness during conditioning; however, only cells that responded to both odour *and* sugar water to begin with showed such enhancement. This is reminiscent of alpha conditioning, and supports the suggestion made earlier that associative learning involves the enhancement of pre-existing neural connections.

Colour learning

Von Frisch first showed that bees could see and learn colours in 1914. He simply trained foragers to find food in a watch glass on a blue piece of paper, and then, while they were away (returning to the hive, unloading sugar solution, and dancing), he set out an array of new papers and watch glasses.

When the bees returned, they had a choice of several different colours or shades of grey, among which was the paper which matched the training colour. The bees landed without hesitation on the correct colour unless it was red, in which case they confused it with grey. Von Frisch also showed that bees can see and learn ultraviolet as a separate colour.

Later workers have greatly expanded our understanding of colour learning. Opfinger (1931) showed that colour is learned on the approach rather than on departure, as would be expected (in retrospect) in classical conditioning, assuming that the taste of sugar water is the US and the colour a CS. Menzel and his associates (1974) have gone so far as to pin the learning down to the last 3 seconds before landing. Moreover, they have shown that there are strong learning biases in colour learning: violet is learned much faster than green and blue–green (90% accuracy in one trial on a two-colour choice versus 65% for blue–green). Since bees have visual receptors tuned to UV, blue, and green, this learning bias is not the trivial result of peripheral sensory effects. After ten trials, however, all colours are learned equally well (95%), although never as well as odours.

8   Shape and pattern learning

It seems reasonable to suppose that honey bees can remember the distinctive shapes and patterns of colours on blossoms, but early work on this question yielded surprising results. Hertz (1929, 1930, 1931) found only the crudest ability to learn shapes; her bees could distinguish a cross from a circle, but not a circle from a square. In addition, she discovered a powerful spontaneous attraction to figures with a high spatial frequency (that is, a high ratio of edge to area); thus, bees were attracted more to a shape with several "petals" than to a circle of the same area, but were attracted most strongly to a checkerboard. Hertz thought that the confusion between a circle and square in her tests as opposed to the learned distinction between a cross and a circle might be explained if all that bees were able to learn about flower shape was the spatial frequency. If this were the case, then this value would be measured and stored as a number rather than as a picture, and consulted, along with colour and odour information, when the bee returned to the field to search.

The idea that bees might store pattern information as a numerical value or *parameter* gained much ground from work in the 1970s (Anderson, 1972, 1977a; Cruse, 1972; Schnetter, 1972; Ronacher, 1979a, 1979b; Wehner, 1972). With the exception of Wehner, each of these researchers concluded that bees remember shape and pattern by means of a list of parameters, with spatial frequency as the most important of the two to six values thought to be

stored; the others include, in decreasing order of presumed importance, relative colour areas (that is, the proportion of different colours incorporated in
the pattern, as when a blossom has 60% of its surface coloured yellow, 30%
blue, and 10% black), the distribution of line angles (that is, the proportion
of different edge angles in the figure), and the distribution of borders (for
example, 8% of the borders oriented at 30° and having blue above and yellow
below). Later, even Wehner (1981) embraced the parameter hypothesis.
However, a close look at the evidence for parameter storage (Gould, 1984a)
reveals many potential flaws in experimental design and interpretation.

One major source of likely difficulty was the implicit assumption that
visual memory in bees, if it existed, would be eidetic: that the resolution of
the memory would correspond to the acuity of real-time visual experience,
which, for bees, is about 1–2°. However, eidetic memory is rare even in
humans (Gleitmann, 1985). If tested with the expectation that bees will remember all that they can resolve, researchers would inevitably conclude that
pattern memory is not pictorial.

A second assumption was that bees remember all that they can about a
pattern; a failure to remember part of a pattern, then, is taken as evidence
against pictorial memory. However, even rats, whose pictorial memory is
well established, remember as little of a pattern as is necessary (Lashley,
1938); only when a similar, unrewarded (or punished) pattern is presented
along with the pattern to be learned do rats use their full capacities. Unrewarded alternative patterns have only rarely been used in work on pattern
and shape learning in bees.

Finally, there is the assumption that the kinds of transformations the nervous system performs on raw sensory data from the eyes is similar in humans
and bees; as a result of this assumption one researcher (Anderson, 1972) performed a series of "generalization" tests. In these experiments, bees were
trained on a figure such as a triangle, and then later offered a choice between
a triangular and a square checkerboard pattern. The bees' failure to show
any preference between them was taken as evidence against pictorial memory
since we, as humans, perceive the triangular checkerboard as in some ways
similar to the training figure. That bees, like humans, should focus their
attention on the outside edge of a figure and largely ignore the interior detail
was simply assumed.

The essential difference between a parameter-based storage system and
pictorial memory is that only the latter necessarily encodes the relative position of the elements of the pattern (Gould, 1984a). By using patterns with
similar parameters but different configurations, and by training in the presence of an unrewarded alternative pattern, usually quite like the reinforced
figure, Gould (1985b, 1986b) demonstrated pictorial memory (Fig. 1). By
increasing the number of elements in the figure (and so reducing the size of

the individual elements) he was also able to estimate the resolution of flower memory to be about 8°. This relatively low resolution, non-eidetic storage at least has the advantage of taking up less neural space than eidetic memory would (16–64 time less), and bees are able to compensate for this deficiency by flying as close to blossoms as they like.

In later work designed to probe the pattern-memory system of bees, Gould (in press a, b) looked for evidence of neural "short cuts" based on the usual radial symmetry and one- or two-colour shading of flowers. He found that bees were perfectly capable of learning asymmetrical, 16-element, four-colour patterns. He also asked whether bees tend to confuse mirror images or rotations of the training patterns, neural transformations that might be adaptive given the usual bilateral symmetry of flowers and the possibility of approaching blossoms from novel perspectives. The only "ambiguity" he found was a willingness to accept a left–right mirror image of the training figure when the alternative was novel (Fig. 2). This is similar to the left–right ambiguity in human infants documented by Bornstein (1979).

Shape is learned more slowly than either odour or colour, and this asymmetry is paralleled by a strong bias in the way these cues are used by honey bees in identifying flowers (Gould, 1984a). For instance, if bees are trained to a blue, triangular, peppermint-scented target, they can, after a few visits, readily distinguish it from an orange, circular, orange-scented target (Fig. 3A). However, if the training cues are separated, and the bees offered a choice between the correct odour combined with incorrect colour and shape on the one hand (i.e. an orange, peppermint-scented circle) versus the correct shape and colour but the wrong odour on the other, they unambiguously choose the alternative with the training odour (Fig. 3B); odour is clearly the most important cue. Only when both alternatives have the correct odour do bees consult the next cue on their flower-identification hierarchy, namely colour (Fig. 3D). When both odour and colour are correct, bees pay attention to shape (Fig. 3E).

9   Landmark learning

Honey bees are able to locate an inconspicuous source of food by triangulation, using nearby landmarks (von Frisch, 1967). On sunny days only one landmark may be needed since the bee can remember and determine the direction from the landmark to the food using the sun as a directional guide; however, in overcast conditions, at least two are required (Hoefer and Lindauer, 1976). It seems likely that this use of landmarks is part of the more general system of landmark navigation employed by bees. Von Frisch and Lindauer (1954) showed that if a hive was established near a prominent

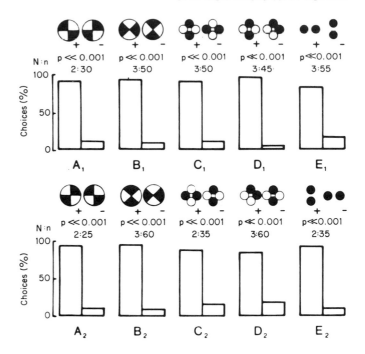

Fig. 1   Bees were trained to find food on the pattern marked "+"; the alternative pattern was unrewarded. During testing, neither pattern offered food. Landings are scored; number of landings, number of bees tested, and *P* values are indicated above the bar graphs. Most experiments were performed in reciprocal pairs, with the reward

landmark, such as a line of trees or a lake shore, and the bees were trained along the landmark to food, the hive could be moved to another tree line or shore and the trained bees would search in the same location relative to the landmark. This was true even if the actual compass direction of the landmarks differed between the two sites. Apparently, conspicuous landmarks come to take precedence over celestial cues in navigation to familiar sites. Later, Dyer and Gould (1981) demonstrated that, in overcast conditions, bees even use conspicuous landmarks to calibrate their navigational compass and, subsequently, their dances.

Traditionally, researchers have assumed that bees (and insects in general) use landmarks in route-specific sets, such that the path to food source A is remembered in terms of a series of landmarks (A_1, A_2, etc.) regularly encountered on the way there (Wehner, 1981). The route to B has its own set of landmarks (B_1, B_2, etc.), again arranged in memory rather like the sequence of frames in a motion picture. This hypothesis is essentially identical to the way

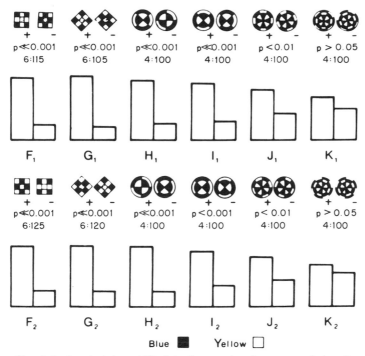

Blue ■   Yellow □

offered during training shifted to the previously unrewarded pattern; these experimental pairs are arranged one above the other in the figure. The data indicate that bees remember floral patterns as pictures.

in which rats were thought to store information used in running mazes as a sequence of cues and responses.

About 40 years ago, Tolman (1948) showed that rats do not normally learn mazes in terms of a route-specific sequence; instead, they form a mental map of the maze, a bird's-eye view, based on their experience in the maze and navigational cues (usually landmarks on the laboratory ceiling) used to judge direction. As a result, they can formulate novel routes when the maze is altered. Tolman called this mental map a "cognitive map", and the term is now used for any mental construct that is used to formulate a novel plan of action independent of physical trial-and-error manipulation. If bees have cognitive maps, they would be able to integrate their knowledge of landmarks encountered on their way to different sites into a map specifying the relative location of all these landmarks; that is, they would know the spatial relationship between A_1 and B_2, even though they had not seen them on the same flight.

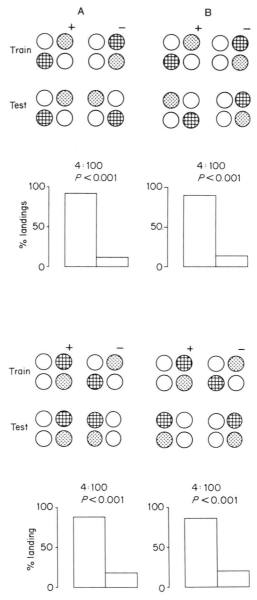

Fig. 2 Bees were trained to find food on the pattern marked "+" in the presence of the unrewarded alternative shown, and then tested with two patterns. In one case, the testing pair included the training pattern and its mirror image, while in the other the mirror image and a novel pattern were offered. The data indicate a partial mirror-image ambiguity.

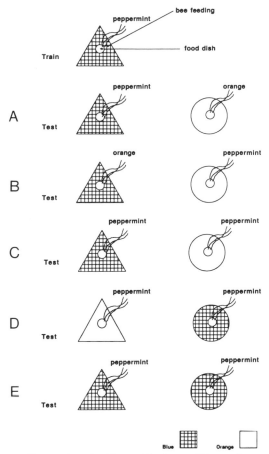

Fig. 3   Experiments illustrating the hierarchical use of learned cues are summarized qualitatively. Training was to a blue, peppermint-scented, triangular-shaped target; testing involved separating the cues between alternative targets to see which are most important. Odour is the most important, followed by colour and then shape.

To test whether landmark knowledge is organized in this way, Gould (1986c) trained bees to one site, and then caught them as they left the hive, displaced them to another location out of sight of their goal, and released them. If bees had only route-specific navigation, they could, at best, recognize the release site as part of a familiar route to some other goal, use that information to navigate back to the hive, and there pick up the landmark trail to their original goal. On the other hand, if bees have cognitive maps, the displaced foragers should be able to place themselves on their mental maps and determine the true direction of the unseen goal. In that case, they should

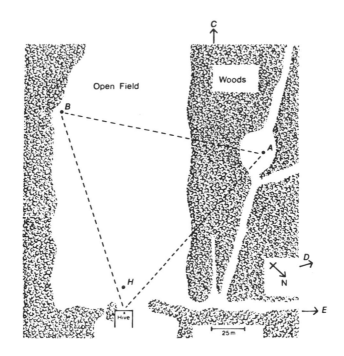

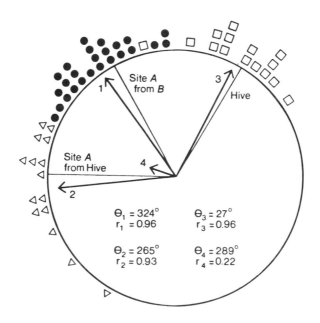

depart directly for the training site along what would be a novel route. This is precisely what the bees did in tests at 160 m (Fig. 4) and 400 m.

The way in which landmark information is used in triangulating an inconspicuous food source has also been a subject of controversy. Anderson (1977b) proposed a very simple, parameter-like system in which bees simply score the presence or absence of landmarks in different visual "sectors" around the source. This ingenious hypothesis was developed to account for some very striking data: when bees were trained to find food at a spot surrounded by landmarks (Fig. 5A), and some of those landmarks were subsequently removed, they then searched closer to the remaining landmarks than would be expected if the bees had simply stored a picture of the landmarks and tried to match that photograph with the landmarks that were left (Fig. 5B).

Later, Cartwright and Collett (1982, 1983) suggested that bees have an eidetic memory of landmarks. In most of their experiments, the bees were trained on three landmarks located to one side of the food (Fig. 5C) and tested with the set altered in such a way that still left a location with which the returning bees could make an approximate visual match (Fig. 5D), and it was there that the bees searched. It is not obvious that these experiments demonstrate that landmark memory is even pictorial, much less eidetic; Anderson's sector-occupancy hypothesis predicts the same results, as long as there are enough sectors.

Gould (1985a, 1986a, 1987) has shown conclusively that bees remember the shape and colour of landmarks photographically (Fig. 6) with a resolution of about 3·5°. Moreover, he established that the reason why Anderson's bees searched closer to the remaining landmarks was that they confused the remaining landmarks with some of the missing ones, reflecting a strategy of using the "outliers" in triangulation. When Gould repeated Anderson's

---

Fig. 4   Bees were trained to find food at a Site A. Later, the same bees were caught on their departure from the hive to Site A and transported in the dark to Site B, where they were released and tracked. These bees departed directly for Site A, even though it was not visible from B (filled circles on the periphery of the circular distribution; each circle represents the departure bearing of one bee; the solid line to the periphery indicates the true direction, while the arrow is the mean vector for the bearings). These data indicate that bees have true locale maps. As controls, the same bees were captured on departure from the hive, transported halfway to Site B and then back to near the hive (Site H), where they were released; their bearings (open triangles) represent the direction expected if the bees displaced to Site B had simply adopted their usual hive-departure bearing. Other bees, not trained to Site A, were captured returning to the hive and transported to Site B where they were released; their bearings (open squares) represent the results expected if the trained bees displaced to Site B had needed to first return to the hive to find their way to Site A.

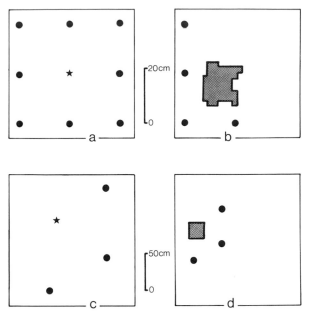

Fig. 5   In Anderson's experiments, bees learned to locate an inconspicuous food source surrounded by landmarks (*a*). When some of the landmarks were removed, the bees searched closer to the remaining landmarks, as indicated by the shading (*b*). In Cartwright and Collett's experiments, three landmarks, located on one side of the food, were available (*c*). During testing (*d*), all the landmarks remained, but were altered in some simple way such that there was still a location from which searching bees could see each of the landmarks in the same relative locations experienced during training.

experiment with distinctive landmarks, the bees showed no such confusion. The probable logic behind the higher resolution in the non-eidetic storage of landmark information as compared with flower memory is that bees must be able to resolve the landmarks while actually at the food site; thus, flying closer to them is of no use if that takes the bee away from the food.

Landmarks are less important than odour and colour in the flower-identification hierarchy of bees, but the relationship between shape and landmarks varies with race (Lauer and Lindauer, 1971). In *carnica* landmarks are more important than shape, while for *ligustica* the order of importance is reversed.

10   Time learning and the organization of memory

As von Frisch (1967) pointed out, different species of flowers provide nectar

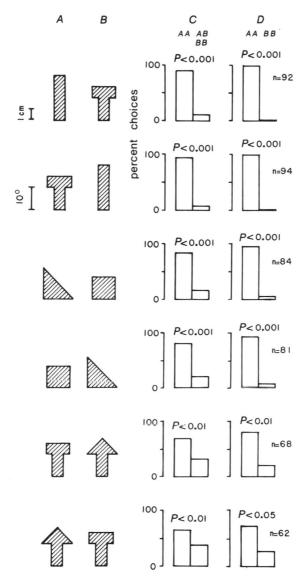

Fig. 6 Honey bees were trained to find an inconspicuous food source using nearby landmarks of the shape indicated by the "+" symbol. During testing, two identical arrays were available, one incorporating the training landmarks and the other employing novel shapes. Bees were able to select the training shape, indicating that the landmark memory is pictorial.

at different times of day. It came as no surprise, therefore, when he discovered that bees learn the time of day to visit particular flowers. Koltermann (1974) succeeded in training individual bees to visit nine different sources over the course of a day, remembering each separately and returning to it at the appropriate time on subsequent days. Honey bees may be able to learn more than nine sources; no one has yet established the limit. The resolution of time memory in Koltermann's tests was about 20 minutes.

Although bees learn the various characteristics of flowers at different rates and subsequently use them in a hierarchical fashion, they appear to store all the information as a unit (Bogdany, 1978). The key feature in this storage system appears to be time of day. Hence, it is impossible to teach bees that two sources are available simultaneously; they can remember one, or the other, or only what the two have in common. The storage system may be compared to an appointment book, with only one "line" for each unit of the day. This line is used for entering odour, colour, shape, and so on. This organization is consistent with several unusual aspects of honey bee behaviour. For example, if a group of bees is trained to collect food from a blue, triangular, peppermint-scented target from 9 o'clock to 10 o'clock, and a yellow, circular, lemon-scented target from 10 o'clock to 11 o'clock, on subsequent days they will be faithful to these food sources even if they are side by side and both available all morning. The bees will begin arriving about 8:45 and land exclusively on the blue target until about 9:45. At this point some will begin to abandon the blue target for the yellow one, even though the blue source continues to provide food; by 10:15 all will have switched targets. It is as though they are compelled by their mental appointment books to move on to the next commitment. Similarly, when one cue associated with a flower is changed (the odour, for example), the bees must relearn all the others. Apparently, changing one part of an entry necessitates erasing the entire set of notes. Adding a previously missing cue, on the other hand, does not affect the information already learned.

It seems clear that the mind of the honey bee is prewired to recognize, learn, store and use the relevant cues according to a pre-ordained weighting. This weighting—the hierarchy—seems sensible given the natural history of bees. Odour is the most constant and predictive cue associated with a flower. Colour is less constant, because of both natural variation and differences in illumination in the field; shape is even less predictive. Prewiring the acquisition means that the insect does not need to discover which cues are predictive, nor to what degree; the bee does not need to realize that a time-exclusive storage system is the most sensible, since flowers are time constant and there can be only one best choice at any given time of day.

11   Learning to handle flowers through trial and error

Because bees quickly memorize the odour, colour, shape, location relative to landmarks, and temporal characteristics of each food source they visit, they are quickly able to relocate rewarding blossoms on subsequent foraging bouts. All of this learning involves the programmed formation of Pavlovian (stimulus–stimulus) associations in which the CSs are further linked to each other in discrete, hierarchically-organized sets, one set for each time of day and flower species. However, these stimulus–stimulus associations are not all that must be learned. There are many different flower morphologies, and most require different strategies for harvesting the reward, a problem which the bees must solve through trial-and-error learning.

As first pointed out by Darwin (1876; cited in Waser, 1986), the necessity of learning how best to manipulate each flower type probably explains why individual flower-feeding insects tend to be species-constant in their foraging. An insect which has recently invested its time and energy learning how to handle a particular species of flower will forage more efficiently if it limits itself to the species that it "knows"; other species will be worked more slowly and clumsily. In addition, the learning itself may proceed more quickly if the insect works only one type of flower at a time (Waser, 1986).

A few observations on bumble bees and butterflies support these ideas. Inexperienced bumble bees spend more time and make more errors per flower visit than experienced ones, and the more complex the flower, the greater the experience needed to develop proficiency (Heinrich, 1976, 1979a, 1979b; Laverty, 1980). Similar phenomena have been observed in the cabbage butterfly, *Pieris papae*; the time which elapsed between the butterfly's landing on a flower and finding the nectar decreased from over 10 seconds to about 2 seconds after a few visits. Furthermore, the learning was fastest when the butterflies visited only one species of flower at a time (Lewis, 1986).

Although it seems clear that these insects are learning something, we know nothing about exactly what is learned or how the memory is stored. Can specific flower-handling behaviours be operantly conditioned experimentally? Are the flower-handling motor patterns stored in a time-linked fashion like the associative memory of the other floral cues? Is the flower-handling memory incorporated into the hierarchically organized memory set for each flower type, so that changing the odour of the food, for example, erases the flower-handling memory as it does the colour memory? Finally, do bees have an innate and limited set of flower-handling motor subunits which are linked together through trial-and-error to yield more complex and functional "crystallized" motor patterns?

Gould (in press) began to address some of these issues experimentally. Individually marked bees were trained to collect sugar water from an

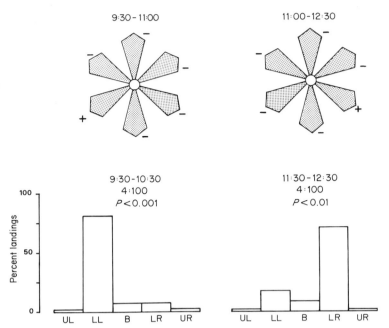

Fig. 7 Honey bees were trained to land only on the lower left petal of an artificial flower in the first part of the morning, and then on the lower right petal during the second part. On subsequent days, the bees preferred to land on the petal that they had been trained on during the same parts of the morning. These data indicate that bees can remember how to approach specific flowers, and store this memory in relation to the time of day.

artificial flower with six "petals", each of which could be individually flicked by a solenoid mounted behind it. Training consisted of allowing the bees to land on only one particular petal; if they landed on one of the five incorrect petals they were punished by being flicked off. This task was easily mastered by the bees; they evidently can and do learn how to land on flowers. A second experiment reveals something of how the memory is stored. In this experiment, the bees were trained to land on the lower left petal from 9:30–11:00 am, and then on the lower right petal of the same flower from 11:00–12:30. If the bees remember how to land according to the colour, shape, or location of the flower, then during testing, in the absence of any reward or punishment, on subsequent days they would be expected to choose the last petal they were trained on—the lower right one—whereas if the memory is time-linked, like the associative memory of the other floral cues, they would be expected to choose on the basis of which petal was correct at that time of day during training. The results, shown in Figure 7, clearly indicate that the memory is stored according to time of day. Presumably, the other details of flower handling are stored in the same way. This is a start; further experiments of this type should help to answer some of the many other questions about trial-and-error learning of flower handling in bees.

12   Learning as a tool in the sensory biology of bees

It is well known that when von Frisch (1914) first showed that bees can learn to associate colours with food, he was more interested in the bees' colour vision than in their learning capacity. He trained the bees simply because they showed no obvious and easily measurable spontaneous response to colours, which is why people supposed they were colour-blind in the first place. Ever since those early experiments, conditioning has been perhaps the single most powerful tool in the hands of biologists studying the sensory capabilities of animals. The colour vision and olfactory sense of bees have been thoroughly analysed by von Frisch and his descendants through conditioning experiments (von Frisch, 1967; Menzel, 1979). Thanks mainly to these studies, the senses of honey bees are now better understood than those of any other invertebrate, and the analysis continues. Here we will limit ourselves to a few particularly interesting and recent examples.

Menzel (1981) trained walking bees to turn right or left in a T-maze according to a visual stimulus presented at the intersection. The bees had to turn one way to obtain food if the visual stimulus was a monochromatic (single wavelength) light and the other way if it was achromatic (white, which is a mixture of wavelengths). When the trained bees were tested with varying intensities of the monochromatic light, Menzel found that at low intensities the bees treated it as if it were achromatic and turned the wrong way. At higher intensities they turned the opposite way, as they were trained. This shows that bees have a visual threshold at which the presence of light, but not its colour, is detected. Such achromatic vision at low light intensities also occurs in vertebrates.

In other experiments on the bees' colour vision, Srinivasan and Lehrer (1985) trained free-flying bees to choose between two light stimuli to obtain food. The rewarded stimulus was a homogeneous mixture of two colours, while the unrewarded one was only one of the colours in pure form. In tests, the bees were made to choose between the previously rewarded colour mixture and a light flickering alternately between the two colours at various flicker-frequencies. If the bees could distinguish the two different colours in the flickering stimulus, they were expected to avoid it because it contained the previously unrewarded pure colour. If, however, the flicker-frequency was high enough for the bees to perceive the flickering stimulus as a steady mixture, then the bees would choose randomly because the two stimuli would

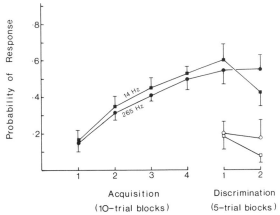

Fig. 8   Two groups of bees were trained to withdraw from a feeder in response to a sound which predicted a mild electric shock. The sound for one group was 265 Hz (the fundamental frequency in the bees' dance sound), and for the other group, 14 Hz (the frequency at which dancing bees waggle). Most bees in each group responded with a probability of about 50% after 40 training trials (acquisition, left; note that the abscissa is marked in ten-trial blocks). After the bees in each group were trained in this way, they were given an additional ten training trials interspersed with ten test trials using the sound on which they had not been trained (discrimination, right; note that the abscissa is marked in five-trial blocks). The bees continued to respond to the original training frequency (solid symbols), but much less to the untrained test frequency (open symbols). Bars show standard errors ($n = 8$ bees in each group). The results indicate that bees can detect both frequencies and, furthermore, can discriminate between the two frequencies.

appear the same. In this way Srinivasan and Lehrer were able to measure the temporal resolution of the honey bees' colour vision. It is about 100 Hz, roughly twice as fast as our own.

Using a similar discriminative training procedure with free-flying bees, Walker and Bitterman (1985) have recently shown that bees can detect and take note of magnetic stimuli while foraging. The magnetic sense of bees had been discovered and analysed to some extent earlier (Lindauer and Martin, 1968, 1972; Martin and Lindauer, 1977; Gould et al., 1978, 1980; Towne and Gould, 1985), but the conditioning technique is by far the most promising one for future analyses of this recently discovered sense.

Towne (in preparation) has recently shown that bees can be trained to respond to airborne sounds, which involves a previously undemonstrated capacity in bees. Free-flying bees were individually trained to visit a feeder, constructed so that a mild electric shock could be delivered to a feeding bee only when its proboscis was in contact with the sugar water food. Abramson (1986) had shown earlier that bees visiting such a feeder could learn to avoid being shocked by withdrawing from the sugar water in response to a CS alone (a puff of air or substrate vibrations), if the CS predicted the shock. Towne used two different frequencies (14 Hz and 265 Hz) of airborne vibration in CSs in the Abramson procedure, and the bees eventually learned to respond to each (Fig. 8, left). In addition, bees that were trained to respond to one of these frequencies also responded, although to a much lesser extent, to the other frequency, indicating that the two are perceived as different by the bees (Fig. 8, right). This acoustic sense, which probably involves the detection of air particle movements rather than pressure oscillations, is particularly interesting because it seems likely to be critically involved in the transfer of information between dancers and recruits in the bees' well-known dance-language communication system (Towne, 1985; Michelsen et al., 1986). (The frequencies of 14 Hz and 265 Hz were chosen for these experiments because they are both produced by dancing bees.) How bees detect these airborne vibrations, what other frequencies (if any) they can detect, and whether or not this sense is indeed used in the dance communication are still not known, but further conditioning experiments should yield some answers.

13   Conclusions

Honey bees face the problem of learning to recognize and handle flowers with a useful set of predispositions and programs: they recognize flower-like targets innately, land on them, explore them and, if they provide food, learn about them. Bees learn both to recognize and to handle flowers. Recognition learning is a case of classical conditioning in which a compound CS is committed to memory. Like many—probably most—cases of learning, the memorizing process is highly biased, with some cues learned more easily than others, and some that are easily perceived not learned at all (polarization, for example). Even within a modality, some colours, odours, and shapes are learned more easily than others. Interestingly enough, the modalities are used in a hierarchical fashion in later flower identification, in an order that corresponds to the speed of initial learning. Similarly, the spontaneous biases within modalities which cause bees to prefer to explore certain colours, shapes, and odours correspond to the biases in learning within that modality. Bees store flower memory in a time-linked, time-exclusive set, in a mental array that must be prewired. These observations, taken together, suggest that honey bee learning is organized in a manner very similar or identical to the circuitry that underlies alpha conditioning: the inputs to the cell(s) responsible for storage—cells with specific time constraints in the case of bees—come from a limited set of modalities, some more strongly represented, and thus more easily learned, than others. Within modalities, some inputs are more strongly represented than others. The result is that certain characteristics, features which are most likely to correspond to flowers, are most readily learned, while the capacity for learning about even very implausible flowers is nevertheless retained. The advantages of organizing learning in this way when the situation is predictably unpredictable are clear, and the same design logic may have evolved in other creatures facing similar challenges.

References

Abramson, C. I. (1986). Aversive conditioning in honeybees (*Apis mellifera*). *J. comp. Psychol.* **100**, 108–116.

Anderson, A. M. (1972). The ability of honey bees to generalize visual stimuli. *In* "Information Processing in the Visual System of Arthropods" (Ed. R. Wehner), pp. 207–212. Springer Verlag, Berlin.

Anderson, A. M. (1977a). Shape perception in the honey bee. *Anim. Behav.* **25**, 67–69.

Anderson, A. M. (1977b). A model for landmark learning in the honey bee. *J. comp. Physiol.* **114**, 335–355.

Arduino, P. and Gould, J. L. (1984). Is tonic immobility adaptive? *Anim. Behav.* **32**, 921–922.

Bitterman, M. E., Menzel, R., Feitz, A. and Schafer, S. (1983). Classical conditioning of proboscis extension in honeybees (*Apis mellifera*) *J. comp. Psychol.* **97**, 107–119.

Bogdany, F. J. (1978). Linking of learning signals in honey bee orientation. *Behav. Ecol. Sociobiol.* **3**, 323–336.

Bornstein, M. H. (1979). Perceptual development: stability and change in feature perception. *In* "Psychological Development from Infancy" (Ed. M. H. Bornstein and W. Kessen), Erlbaum, Hillside, New Jersey.

Breland, K. and Breland, M. (1961). The misbehavior of organisms. *Amer. Psychol.* **16**, 681–684.

Brines, M. L. and Gould, J. L. (1979). Bees have rules. *Science* **206**, 571–573.

Carew, T. J., Abrams, T. W., Hawkins, R. D. and Kandel, E. R. (1984). The use of simple invertebrate systems to explore psychological issues related to associative learning. *In* "Primary Neural Substrates of Learning and Behavioral Change" (Ed. D. L. Alkon and J. Farley), pp. 169–184. Cambridge University Press, Cambridge.

Cartwright, B. A. and Collett, T. A. (1982). How honey bees use landmarks to guide their return to a food source. *Nature* **295**, 560–564.

Cartwright, B. A. and Collett, T. A. (1983). Landmark learning in bees. *J. comp. Physiol.* **151**, 521–543.

Connolly, K. (1977). Factors influencing the learning of manual skills by young children. *In* "Constraints on Learning" (Eds R. A. Hinde and J. Stevenson-Hinde), pp. 337–366. Academic Press, London.

Couvillon, P. A. and Bitterman, M. E. (1980). Some phenomena of associative learning in honeybees. *J. comp. Physiol. Psychol.* **94**, 878–885.

Couvillon, P. A. and Bitterman, M. E. (1982). Compound conditioning in honeybees. *J. comp. Physiol. Psychol.* **96**, 192–199.

Couvillon, P. A. and Bitterman, M. E. (1984). The overlearning–extinction effect and successive negative contrast in honeybees (*Apis mellifera*). *J. comp. Psychol.* **98**, 100–109.

Couvillon, P. A., Klosterhalfen, S. and Bitterman, M. E. (1983). Analysis of overshadowing in honeybees. *J. comp. Psychol.* **97**, 154–166.

Cruse, H. (1972). A qualitative model for pattern discrimination in the honey bee. *In* "Information Processing in the Visual System of Arthropods" (Ed. R. Wehner), pp. 201–206. Springer-Verlag, Berlin.

Curio, E. and Vieth, W. (1978). Cultural transmission of enemy recognition. *Science* **202**, 899–901.

Darwin, C. (1876). "On the Effects of Cross and Self Fertilisation in the Vegetable Kingdom". John Murray, London.

Dyer, F. C. and Gould, J. L. (1981). Honey bee orientation: a backup system for cloudy days. *Science* **214**, 1041–1042.

Dyer, F. C. and Gould, J. L. (1983). Honey bee navigation. *Amer. Sci.* **71**, 587–597.

Erber, J. (1975a). The dynamics of learning in the honeybee (*Apis mellifica carnica*). I. The time dependence of the choice reaction. *J. comp. Physiol.* **99**, 231–242.

Erber, J. (1975b). The dynamics of learning in the honeybee (*Apis mellifica carnica*). II. Principles of information processing. *J. comp. Physiol.* **99**, 243–255.

Erber, J. (1980). Neural correlates of non-associative and associative learning in the honey bee. *Verh. Deut. Zool. Gesell.* **73**, 250–261.

Erber, J. (1981). Neural correlates of learning in the honeybee. *Trends Neurosci.* **4**, 270–273.

Erber, J. (1983). The search for neural correlates of learning in the honeybee. *In* "Neuroethology and Behavioural Physiology" (Eds F. Huber and H. Markl), pp. 216–230. Springer Verlag, Berlin.

Erber, J. (1984). Response changes of single neurons during learning in the honeybee. *In* "Primary Neural Substrates of Learning and Behavioural Change" (Eds D. L. Alkon and J. Farley), pp. 275–288. Cambridge University Press, Cambridge.

Erber, J., Mashur, Th. and Menzel, R. (1980). Localization of short-term memory in the brain of the bee, *Apis mellifera*. *Physiol. Entomol.* **5**, 343–358.

Foree, D. and LoLordo, V. M. (1973). Attention in the pigeon: the differential effect of food getting vs. shock avoidance procedures. *J. comp. Physiol. Psychol.* **85**, 551–558.

von Frisch, K. (1914). Demonstration von Versuchen zum Nachweis des Farbensinnes bei angeblich total farbenblinden Tieren. *Verh. Deut. Zool. Gesell.* 1914.

Von Frisch, K. (1919). Ueber den Geruchssinn der Bienen und seine blütenbiologische Bedeutung. *Zool. Jb.* (*Physiol.*) **37**, 1–238.

von Frisch, K. (1967). "The Dance Language and Orientation of Bees". Harvard University Press, Cambridge, Massachusetts.

von Frisch, K. and Lindauer, M. (1954). Himmel und Erde in Konkurrenz bei der Orientierung der Bienen. *Naturwissenschaften* **41**, 245–253.

Gleitmann, H. (1985). "Psychology". Nortons, New York.

Goelet, P., Castellici, V. F., Schacher, S. and Kandel, E. R. (1986). The long and the short of long-term memory—a molecular framework. *Nature* **322**, 419–422.

Gould, J. L. (1980). Sun compensation by bees. *Science* **207**, 545–547.

Gould, J. L. (1982). "Ethology: The Mechanisms and Evolution of Behavior". Nortons, New York.

Gould, J. L. (1984a). The natural history of honey bee learning. *In* "The Biology of Learning" (Eds P. Marler and H. Terrace), pp. 149–180. Springer Verlag, Berlin.

Gould, J. L. (1984b). Processing of sun-azimuth information by bees. *Anim. Behav.* **32**, 149–152.

Gould, J. L. (1985a). Honey bee learning and memory. *In* "The Neurobiology of Learning and Memory" (Eds G. Lynch, J. L. McGaugh and N. Weinberger), pp. 193–210. Guilford, New York.

Gould, J. L. (1985b). How do bees remember flower shape? *Science* **227**, 1492–1494.

Gould, J. L. (1986a). The biology of learning. *Ann. Rev. Psychol.* **37**, 163–192.

Gould, J. L. (1986b). Pattern learning by honey bees. *Anim. Behav.* **34**, 990–997.

Gould, J. L. (1986c). The locale map of bees: do insects have cognitive maps? *Science* **232**, 861–863.

Gould, J. L. (1987). Landmark learning by honey bees. *Anim. Behav.* **35**, 26–34.

Gould, J. L. (in press a). The resolution and timing of pattern learning by honey bees. *Anim. Behav.*

Gould, J. L. (in press b). A mirror-image ambiguity in honey bee pattern recognition. *J. Insect Behav.*

Gould, J. L. (in press c). Operant learning by honey bees. *Anim. Behav.*

Gould, J. L. and Gould, C. G. (1982). The insect mind: physics or metaphysics. *In* "Animal Mind—Human Mind" (Ed. D. R. Griffin), pp. 269–298. Springer Verlag, Berlin.

Gould, J. L., Kirschvink, J. L. and Deffeyes, K. S. (1978). Bees have magnetic remanence. *Science* **201**, 1026–1028.

Gould, J. L., Kirschvink, J. L., Deffeyes, K. S. and Brines, M. L. (1980). Orientation of demagnetized bees. *J. exp. Biol.* **86**, 1–8.

Gould, J. L. and Marler, P. (1984). The natural history of learning. *In* "The Biology of Learning" (Eds P. Marler and H. Terrace), pp. 47–74. Springer Verlag, Berlin.

Hawkins, R. D., Abrams, S. W., Carew, T. J. and Kandel, E. R. (1983). A cellular mechanism of classical conditioning in Aplysia: activity-dependent enhancement of presynaptic facilitation. *Science* **219**, 400–404.

Heinrich, B. (1976). The foraging specializations of individual bumblebees. *Ecol. Monogr.* **46**, 105–128.

Heinrich, B. (1979a). "Majoring" and "minoring" by foraging bumblebees, *Bombus vagans*: an experimental analysis. *Ecol.* **60**, 245–255.

Heinrich, B. (1979b). "Bumblebee Economics". Harvard University Press, Cambridge, Massachusetts.

Hertz, M. (1929). Die Organisation des optischen Feldes bei der Biene, I. *Z. vergl. Physiol.* **8**, 693–784.

Hertz, M. (1930). Die Organisation des optischen Feldes bei der Biene, II. *Z. vergl. Physiol.* **11**, 107–145.

Hertz, M. (1931). Die Organisation des optischen Feldes bei der Biene, III. *Z. vergl. Physiol.* **14**, 629–674.

Hoefer, I. and Lindauer, M. (1976). Das Schatten als Hilfsmarke bei der Orientierung der Honigbiene. *J. comp. Physiol.* **112**, 5–18.

Koltermann, R. (1973). Rassen- bzw. artspezifische Duftbewertung bei der Honigbiene und ökologische Adaption. *J. comp. Physiol.* **85**, 327–360.

Koltermann, R, (1974). Periodicity in the activity and learning performance of the honey bee. *In* "Experimental Analysis of Insect Behaviour" (Ed. L. B. Brown), pp. 218–227. Springer Verlag, Berlin.

Lashley, K. S. (1938). Conditioned reactions in the rat. *J. Psychol.* **6**, 311–344.

Lauer, J. and Lindauer, M. (1971). Genetische fixierte Lerndisposition bei der Honigbiene. *In* "Informationsaufnahme und Informationsverarbeitung im lebenden Organismus I". pp. 1–87. Franz Steiner Verlag, Wiesbaden.

Laverty, T. M. (1980). The flower-visiting behavior of bumblebees: floral complexity and learning. *Can. J. Zool.* **58**, 1324–1335.

Lewis, A. C. (1986). Memory constraints and flower choice in *Pieris rapae*. *Science* **232**, 863–865.

Lindauer, M. (1959). Angeborene und erlernte Komponenten in der Sonnenorientierung der Bienen. *Z. vergl. Physiol.* **42**, 43–62.

Lindauer, M. (1976). Recent advances in the orientation and learning of honey bees. *In* "Proceedings of the XV International Congress on Entomology", pp. 450–460. Washington, D.C.

Lindauer, M. and Martin, H. (1968). Die Schwerorientierung der Bienen unter dem Einfluss der Erdmagnetfelds. *Z. vergl. Physiol.* **60**, 219–243.

Lindauer, M. and Martin, H. (1972). Magnetic effects on dancing bees. *In* "Animal Orientation and Navigation" (Eds S. R. Galler, K. Schmidt-Koenig, G. J. Jacobs and R. E. Belleville), pp. 559–567. NASA SP-262, U.S. Govt. Printing Office, Washington, D.C.

LoLordo, V. M. (1979). Selective associations. *In* "Mechanisms of Learning and Motivation" (Eds A. Dickenson and R. A. Boakes). Erlbaum, Hillsdale, New Jersey.

Mackintosh, N. J. (1974). "The Psychology of Animal Learning". Academic Press, New York.

Marler, P. (1984). Song-learning: innate species differences in the learning process. *In* "The Biology of Learning" (Ed. P. Marler and H. S. Terrace), pp. 289–310. Springer Verlag, Berlin.

Martin, H. and Lindauer, M. (1977). Der Einfluss der Erdmagnetfelds und die Schwerorientierung der Honigbiene. *J. comp. Physiol.* **122**, 145–187.

Menzel, R. (1979). Behavioural access to short-term memory in bees. *Nature* **281**, 368–369.

Menzel, R. (1981). Achromatic vision in the honeybee at low light intensities. *J. comp. Physiol.* **141**, 389–393.

Menzel, R. (1983). Neurobiology of learning and memory: the honeybee as a model system. *Naturwissenschaften* **70**, 504–511.

Menzel, R. (1984). Biology of invertebrate learning. *In* "The Biology of Learning" (Eds P. Marler and H. S. Terrace), pp. 249–270. Springer Verlag, Berlin.

Menzel, R. (1985). Learning in honeybees in an ecological and behavioural context. *In* "Experimental Behavioural Ecology and Sociobiology" (Eds B. Holldobler and M. Lindauer), pp. 55–74. Gustav Fischer Verlag, Stuttgart.

Menzel, R. and Bitterman, M. E. (1983). Learning by honeybees in an unnatural situation. *In* "Neuroethology and Behavioural Physiology" (Eds F. Huber and H. Markl), pp. 206–215. Springer Verlag, Berlin.

Menzel, R., Erber, J. and Mashur, Th. (1974). Learning and memory in the honeybee. *In* "Experimental Analysis of Insect Behaviour" (Ed. L. Barton-Browne), pp. 195–217. Springer Verlag, Berlin.

Michelsen, A., Kirchner, W. H. and Lindauer, M. (1986). Sound and vibration signals in the dance language of the honeybee *Apis mellifera*. *Behav. Ecol. Sociobiol.* **18**, 207–212.

Moore, B. I. (1973). Role of directed Pavlovian reactions in simple instrumental learning in the pigeon. *In* "Constraints on Learning" (Eds R. A. Hinde and J. Stevenson-Hinde), pp. 159–188. Academic Press, New York.

Opfinger, E. (1931). Ueber die Orientierung der Biene an der Futterquelle. *Z. vergl. Physiol.* **15**, 431–487.

Pavlov, I. P. (1927). "Conditioned Reflexes". Oxford University Press, Oxford.

Ronacher, B. (1979a). Aequivalenz zwischen Gross- und Helligkeitsunterschieden im Rahmen der visuellen Wahrnehmung der Honigbiene. *Biol. Cyber.* **32**, 63–75.

Ronacher, B. (1979b). Beitrag einzelner Parameter zum wahrnehmungsgemässen

Unterschied von zusammengesetzten Reizen bei der Honigbiene. *Biol. Cyber.* **32,** 77 83.

Sahley, C. L. (1984). Behavior theory and invertebrate learning. *In* "The Biology of Learning" (Eds P. Marler and H. S. Terrace), pp. 181–196. Springer Verlag, Berlin.

Sahley, C. L., Rudy, J. L. and Gelperin, A. (1984). Associative learning in a mollusc: a comparative analysis. *In* "Primary Neural Substrates of Learning and Behavioral Change" (Eds D. L. Alkon and J. Farley), pp. 243–258. Cambridge University Press, Cambridge.

Schnetter, B. (1972). Experiments on pattern discrimination in honey bees. *In* "Information Processing in the Visual System of Arthropods" (Ed. R. Wehner), pp. 195–200. Springer Verlag, Berlin.

Schwartz, B. (1978). "Psychology of Learning and Behavior" (1st edn). W.W. Norton and Co., New York.

Schwartz, B. (1984). "Psychology of Learning and Behavior" (2nd edn). W.W. Norton and Co., New York.

Skinner, B. F. (1938). "Behavior of Organisms". Appleton-Century-Crofts, New York.

Srinivasan, M. and Lehrer, M. (1985). Temporal resolution of color vision in the honeybee. *J. comp. Physiol.* **157,** 579–856.

Staddon, J. E. R. and Simmelhag, V. L. (1971). The "superstition" experiment: a reexamination of its implications for the principles of adaptive behavior. *Psychol. Rev.* **78,** 3–43.

Thorndike, (1898). Animal intelligence: An experimental study of the associative process in animals. *Psych. Monogr.* **2**(8).

Tolman, E. C. (1948). Cognitive maps in rats and men. *Psychol. Rev.* **55,** 189–208.

Towne, W. F. (1985). Acoustic and visual cues in the dances of four honey bee species. *Behav. Ecol. Sociobiol.* **16,** 185–187.

Towne, W. F. and Gould, J. L. (1985). Magnetic field sensitivity in honey bees. *In* "Magnetite Biomineralization and Magnetoreception in Organisms" (Eds J. L. Kirschvink, D. S. Jones and B. J. MacFadden), pp. 385–406. Plenum Press, New York.

Walker, M. W. and Bitterman, M. E. (1985). Conditioned responding to magnetic fields by honeybees. *J. comp. Physiol.* **157,** 67–71.

Waser, N. M. (1986). Flower constancy: definition, cause, and measurement. *Am. Nat.* **127,** 593–603.

Wehner, R. (1972). Pattern modulation and pattern detection in the visual system of Hymenoptera. *In* "Information Processing in the Visual System of Arthropods" (Ed. R. Wehner), pp. 183–194. Springer Verlag, Berlin.

Wehner, R. (1981). Spatial vision in arthropods. *In* "Handbook of Sensory Physiology" (Ed. H. Autrum), pp. 287–616. Springer Verlag, Berlin.

The Formation of a Neurohaemal Organ During Insect Embryogenesis

Paul H. Taghert, Jeffrey N. Carr and John B. Wall

Department of Anatomy and Neurobiology, Washington University School of Medicine, 660 S. Euclid Avenue, St Louis, MO 63110, USA

ADVANCES IN INSECT PHYSIOLOGY VOL. 20
ISBN 0–12–024220–6

1 Introduction

1.1 NEUROENDOCRINE CELLS AND NEUROHAEMAL ORGANS

Neuroendocrine cells are specialized neurons that project axons into peripheral structures, termed neurohaemal organs. Within these organs the axons terminate with a profusion of varicosities and "blind endings" from which neurohormones are released into the blood (Carrow *et al.*, 1984; Copenhaver and Truman, 1986). In this chapter, we consider the embryonic development of a segmentally-repeated neurohaemal organ—the transverse nerve—in the tobacco hornworn, *Manduca sexta*. This cellular analysis of developing neurosecretory neurons has as its goals to understand (1) the stereotyped formation of axonal projections, (2) the timing and induction of varicosity formation along terminal branches, (3) the basis of cell-specific expression of neurotransmitters and neurohormones at developmentally relevant times.

During the embryonic development of the nervous system, individual neurons differentiate along highly stereotyped lines. Morphological growth is accomplished by the elaboration of growth cones which demonstrate precise and cell-specific navigation (Bastiani *et al.*, 1985; Myers *et al.*, 1986). With equal precision, individual neurons begin to express neurotransmitters according to patterns and schedules that are also cell-specific and that in some animals are related to cell lineages (Sulston and Horvitz, 1977; Blair, 1982; Taghert and Goodman, 1984). Two major questions arise in considering neuronal development: how are precise yet diverse cellular phenotypes allotted to individual neurons, and how is the expression of numerous phenotypes smoothly integrated during the course of development for a single cell? In recent years, much progress has been made in the analysis of neuronal determination and differentiation by utilizing insect embryos (Bastiani *et al.*, 1985; Thomas *et al.*, 1984; Taghert *et al.*, 1986). In this chapter, we summarize recent observations we have made concerning the differentiation of a neurohaemal organ during embryonic development in the tobacco hornworm, *Manduca sexta*.

1.2 TRANSVERSE NERVES ARE SEGMENTAL NEUROHAEMAL ORGANS

Each abdominal ganglion of the ventral nerve cord of larval *Manduca* consists of approximately 700 neurons (Taylor and Truman, 1974). Within these segments there are three nerves that emanate from each ganglion to supply muscles and organs and to receive sensory information. The dorsal (DN) and ventral (VN) nerves are bilaterally paired. The third nerve is unpaired and is called the transverse nerve (TN); it originates from the small median nerve that lies between the longitudinal connectives. The TN has long been known to have two main functions in insects: supplying the motor innervation to the closer and/or opener muscles of the spiracles (Lewis et al., 1973); and acting as a point of release for various neurohormones (Truman, 1973; Taghert and Truman, 1982; see Raabe, 1982, for a review).

The peri-visceral organ (PVO) is often used to refer to thickened portions of the TN that lie on each side of the ganglion; the relative length of the PVO in relation to the TN varies between the different insect orders (Raabe et al., 1974). While the PVO is obviously of major importance, it is clear that the neurohaemal function is distributed along the entire length of the nerve because the neurosecretory neurons make endings along its entire length (Fig. 1). In addition, there is physiological evidence of a widespread neuro-haemal function. For example, bursicon activity is distributed throughout the length of the nerve prior to normal hormone release but following release the hormone titre is lowered throughout its length.

In cross-sections, light and electron micrographs reveal a pair of axons that run in an uninterrupted fashion through the TN; by reconstruction of serial images, and from physiological experiments (Taghert, unpublished results), these axons can be identified as the spiracular closer muscle moto-neurons (SP MNs). The MN axons are surrounded by a thick sheath which is composed of many glial cells and their wrapping processes. The other neural components of the nerve are neuroendocrine (NE) cell axons and terminals. These are also ensheathed by glial elements and are distributed throughout the body of the nerve in dispersed yet intercalated units (Fig. 1). More than one glial cell can be associated with an NE axon, and more than one type of NE axon (as judged by the morphology of neurosecretory granules) can be associated with an ensheathing glial group (Taghert, unpublished results). NE elements are separate from the MN unit at all positions along the TN. Release of neurohormones is thought to occur at swellings along the NE axons. At these points the glial processes are no longer closely apposed to them (Fig. 1).

The TN/PVO has been described as releasing numerous biologically-active substances (Raabe, 1982). The two best studied examples are the tanning hormone, bursicon (Truman, 1973; Taghert and Truman, 1982) and the small cardioactive peptides (Tublitz and Truman, 1985) in Manduca. In both instances, the neuropeptides have been localized to specific neurons within the segmental ganglia and were shown to be released in pulsatile fashion within 2 minutes of the onset of wing spreading behaviour in the adult. Experiments that consider the synaptic inputs onto these and other TN-projecting NE neurons of the abdominal ganglia indicate that all these

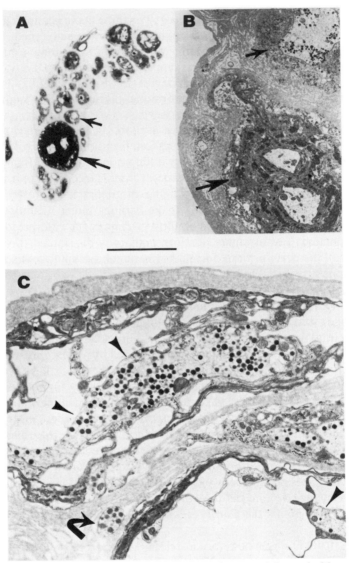

Fig. 1 (A) One micron section of a TN from a pharate adult moth. Note the large central cylinder (large arrow) that is composed of glial cells and the axons of the two MNs. Surrounding the cylinder are smaller inclusions (small arrow) that represent the NE axons that are similarly surrounded by glia. (Scale bar: 25 μ). (B) Electron micrograph of a similar position along the pharate adult TN. MN axons are prominent, as are nearby NE axons that contain large dense cored granules. (Scale bar: 4 μ). (C) Electron micrograph of NE processes in the pharate adult nerve that are mostly free of closely apposed glia (arrow heads) or are coursing freely (curved arrow) through the extracellular matrix. These may represent neurohaemal sites. (Scale bar: 2·5 μ.)

neurons may be under coordinated synaptic control and thus may all release
their substances simultaneously (Taghert, 1981).

1.3   THE TN CONTAINS AXONS OF BOTH CENTRAL AND PERIPHERAL NEURONS

Over the past 10 years, we have compiled a catalogue of all the neurons that
project axons into the TN of abdominal segments in *Manduca*. A certain
number of neurons differentiate during embryogenesis, but this number increases during larval life and again during metamorphosis. By the end of
embryogenesis, there are two motoneurons (the SP MNs), ten central NE
neurons (eight of which produce bursicon), and four peripheral NE neurons
(two of which produce cardioactive peptides). During larval life, two additional central NE neurons are added, and during adult development, six
more central NE neurons appear (Taghert and Truman, 1980). Thus, there
are finally 24 neurons with axons in the TN: two MNs, four peripheral NE
neurons and 18 central NE neurons. Figure 2 shows the disposition of the
TN-projecting neurons following the completion of embryogenesis.

These various neurons take five different routes in projecting axons to the
TN (Fig. 2). The MNs reach the TN via the median nerve that descends from
the ganglion anterior to the TN. Six of the eight bursicon cells and a pair of
NE cells at the ganglion midline project out from the VN and reach the TN
via an anastomosis in the periphery; one pair of bursicon cells project out of
the DN of the next posterior ganglion and reach the TN via a separate anastomosis; finally the peripheral NE cells form their own distinct axon pathway
to the TN.

It is clear from the above description that this simple neuroendocrine effector organ is complex when considered in fine detail. The different individual
neurons are specialized to secrete a variety of hormonal substances and,
further, they reach their end organ (the TN) via diverse routes and at diverse
times. The following sections summarize our recent findings as regards the
embryonic development of this specialized set of neurons.

1.4   EMBRYOGENESIS IN *MANDUCA*

Under laboratory conditions (25°C) *Manduca* embryogenesis is completed
within approximately 100 hours. The external morphology of developing embryos can be used to accurately stage animals to within 3 hours. We use the
convention adopted by Tyrer (1970) and Bentley *et al.* (1979) in staging
grasshopper embryogenesis and refer to *Manduca* embryonic development in
terms of percentages, where 100% is a fully completed embryo at the hatch-
ing stage. Conveniently, 1% of developmental time = 1 hour of real time. The
timetable of *Manduca* embryonic development was initiated by N. Tublitz

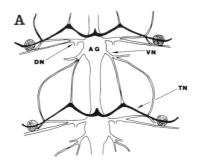

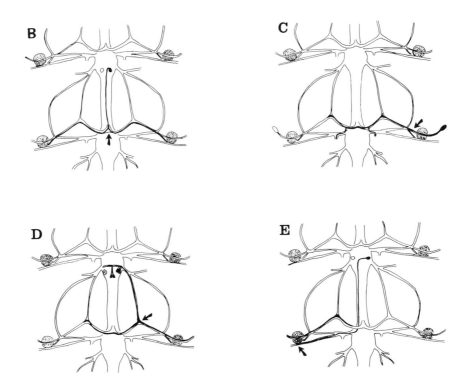

Fig. 2 Axon pathways of the TN-projecting neurons. (A) Diagram showing ideal-
ized arrangement of nerves and ganglia in two abdominal segments. B–E The cell
body positions and axon pathways of the various neurons that project to the TN. (B)
the SP MN's; (C) the peripheral NE cell L1; (D) The central NE cells that exit via the
VN; (E) the central NE cell that exists via the DN in the posterior segment. Single
arrows indicate the points at which the various neurons join the TN. TN: transverse
nerve; DN: dorsal nerve; VN: ventral nerve; AG: abdominal ganglion. Shaded circles
represent spiracles; contralateral neuronal homologues are indicated by open circles.

and M. Bate (personal communication) and has been supplemented by our own observations.

In moth embryos, as in grasshoppers, most neurons of the segmental ganglia are the progeny of a specialized class of cells termed neuroblasts (Bate, 1976; Thomas *et al.*, 1984; Doe and Goodman, 1985). Neuronal progeny are first produced at about the 25% stage and continue to be produced through the late stages of embryogenesis. The results presented here are derived from the observations of living embryos dissected out of the eggshell and embryonic membranes, and viewed in a compound microscope modified for Nomarski interference contrast optics. In addition, whole embryos dissected as above were stained as wholemounts using a variety of serum and monoclonal antibodies. Finally, single cells in living or paraformaldehyde-fixed embryos were dye-filled with microelectrodes containing Lucifer Yellow, and processed with an anti-Lucifer Yellow antiserum (Taghert *et al.*, 1982). Embryo dissections typically flatten the embryos in such a way that the developing ganglia of the ventral nerve cord are in the middle of the tissue and the rest of the body is a two dimensional sheet. The photographs and drawings used here utilize this artificially "filletted" preparation.

1.5   MONOCLONAL ANTIBODIES TO THE TN

In order to examine the development and function of the NE neurons within the TN, we made monoclonal antibodies (Mabs) to individually dissected TNs of pharate adult *Manduca* and selected these antibodies according to their cell-specific patterns of expression within the CNS (Taghert *et al.*, 1983). When screened on embryonic *Manduca*, we found that many of these Mabs recognized what might be termed early differentiation antigens and that they highlighted previously undescribed aspects of neural and non-neural differentiation. We used these antibodies (specifically TN-1) to complement microelectrode dye-filling studies of the embryonic formation of the TN.

2   Formation of the TN

2.1   NON-NEURAL CELLS BUILD A FRAMEWORK

2.1.1   *Strap cells prefigure the TN*

Like the other nerves in this animal, the TN has a characteristic position, shape, and trajectory. These stereotyped features are the result of stereotyped morphogenetic events that begin to occur just as the first neurons are born.

Interestingly, much of the overall detail concerning future TN position and trajectory appears to be specified by a collection of non-neural cells. These cells are probably mesodermal in origin; many of them appear to differentiate into glia that ensheath peripheral nerves.

By 27% the first growth cones are produced by the first born neurons in locusts (Bate and Grunwald, 1981) and in moths (Thomas et al., 1984; Carr and Taghert, unpublished results). At about this time, a group of cells appear, which lie over the anterior portion of each segmental ganglion and are connected on either side to a mass of undifferentiated mesoderm (Fig. 3A). This collection of cells numbers about 50 and, following development to 34%, has changed its appearance. It thins out so that it is only one to two cells wide and is now inserted into the ectoderm at the segment border on either side of each segmental ganglion (Fig. 3B). Moving distally from the insertion point, the group forms a partial ring around the presumptive spiracle. Neuronal growth cones have not yet reached this structure (see below), yet many of the features of the axonal pathways they will form are already in place. We call this group of cells over the ganglion the "strap" and the group of cells around the forming spiracle the "bridge". In Figure 3, these cells are highlighted by means of antibody-staining with Mab TN-1.

By 39%, the strap has taken on an appearance more similar to the mature TN (Fig. 3C); it has thinned out even more, so that individual cells now only occur at intervals along its length. In addition, the axons of the motoneurons can be seen staining and branching at the origin of the TN, to supply both sides of the body. The ganglion has also taken on a mature appearance; prominent axon tracts that run longitudinally are stained by the antibody. Note the absence of staining in most cell bodies, in the neuropil and in the majority of commissural axon bundles (commisures run across the midline of each ganglion, orthogonal to the connectives). By 61%, the TN is nearly fully formed (Fig. 3E); all its component neurons have arrived onto it and have nearly completed their morphogenesis. In addition, cell-specific transmitters have begun to be expressed. The ganglion is larger by the addition of new neurons and by the increase in neuropil size due to the elaboration of the central projections.

This pattern of TN construction is outlined from camera lucida drawings in Figure 4. The developing strap and bridge can be seen in relation to the ganglion, and also seen are the points of arrival of certain neuronal growth cones; these are described more fully below. Next, we discuss the development of the bridge and make an inference as to its role in TN formation.

2.1.2   *Bridge cells prefigure a nerve anastomosis*

As diagrammed in Figure 5, the bridge is a collection of approximately 10–15

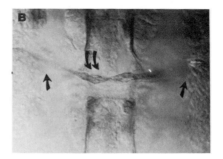

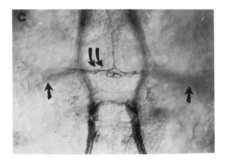

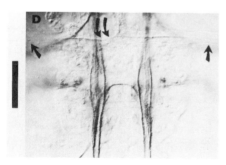
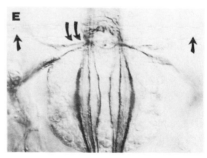

Fig. 3 Photomicrographs of the developing TN as stained with Mab TN-1. (A) 27% of development; (B) 34%; (C) 39%; (D) 49%; (E) 61%. Note that the TN (double arrow) develops as a "strap" of cells at the anterior margin of the forming ganglion. The strap insertion point is marked on either side with curved arrows; in most panels it is out of the plane of focus. (Scale bar: 50 μ.)

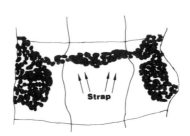

A

Strap

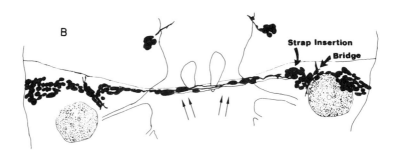

B

Strap Insertion

Bridge

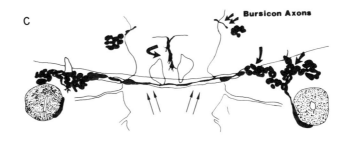

C

Bursicon Axons

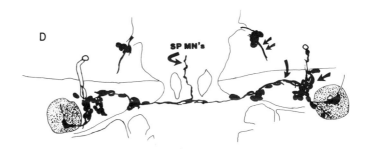

D

SP MN's

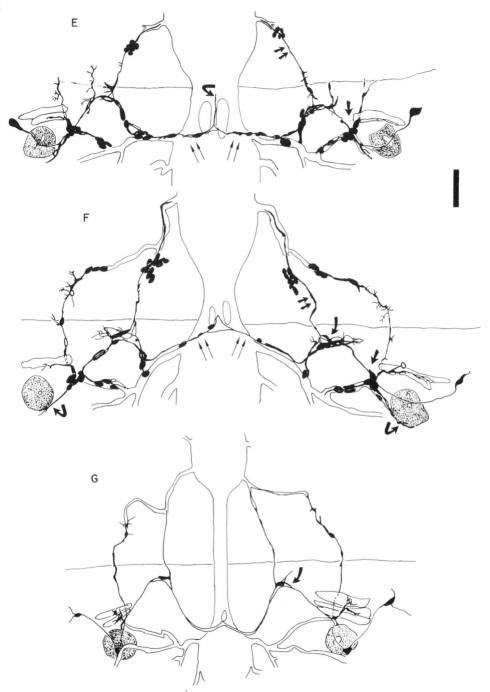

Fig. 4 Camera lucida drawings of the developing TN in abdominal segments. (A) 32% of development; (B) 37%; (C) 43%; (D) 49%; (E) 57%; (F) 61%. (Scale bar: 50 μ.) (G) 90% (Scale bar: 100 μ.) See text for further details.

cells that were once continuous with the strap up to about 39% stage. In mature larvae (Fig. 5F), this structure is an anastomosis between many convergent peripheral nerves. These include (1) the SP MNs from the strap insertion point through the bridge to the muscle target, (2) a nerve connection from the bridge to a branch of the VN from the next anterior segment, and (3) a nerve connecting DN MNs through the bridge to a muscle group that lies above the spiracle.

Interestingly, as visualized by TN-1 immunostaining, this future interchange of axon pathways is already in place by the collection of bridge cells and by their elaboration of processes. We believe that most of these bridge cells later become glial cells (like the strap cells), because throughout embryonic development we observe TN-1 positive cells at these locations and they ultimately come to ensheath the axons that grow along them.

Thus, we believe that two major structures, the *strap* and the *bridge*, provide much of the information necessary to build the TN according to its proper position and trajectory. Future experiments will test the hypothesis that these collections of future glial cells are required for proper TN formation by eliminating them prior to the arrival of specific neuronal growth cones. Although not our primary focus, we have also observed that cells similar in morphology (but not always in antigenic properties) are placed along the pathways of future DN and VN nerve formation (Carr and Taghert, unpublished results). In grasshoppers, specific glial cells are necessary for the proper formation of the intersegmental peripheral nerve (M. Bastiani and C. Goodman, personal communication). In developing adult moths, glial cells appear to be necessary for the proper formation of antennal glomeruli in the brain (Oland and Tolbert, 1986). In addition, non-neural cells (glia) help define future nerve pathways in vertebrate embryos as well (Carpenter and Hollyday, 1986). Thus, the roles of glial cells during development may be more diverse than previously thought.

2.2   NEURAL CELLS GROW ALONG THE NON-NEURAL FRAMEWORK

2.2.1   *TN neurons arrive and grow in stereotyped fashion*

There are basically three groups of neurons that project axons ino the TN during embryonic development: (1) The SP MNs and (2) the peripheral NE's arrive at approximately the same time but from different locations; (3) a group of NE cells (these include the bursicon neurons) arrive slightly later and from a different trajectory. First we will describe this pattern of neuronal differentiation in terms of groups of axons growing roughly in concert to build the nerve proper. Next, we will consider the differentiation of a single

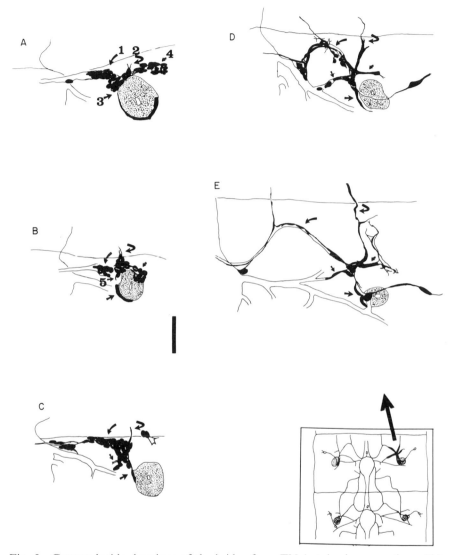

Fig. 5 Camera lucida drawings of the bridge from TN-1 stained preparations. (A) 39% of development; (B) 41%; (C) 43%; (D) 57%. (Scale bar: 50 μ.) (E) 90%. (Scale bar: 67 μ.) Arrows indicate specific groups of stained cells that later became specific nerve branches.

neuron, the peripheral NE cell L1, in order to evaluate the mode and potential mechanisms that are utilized to elaborate a specific neuroendocrine neuron.

2.2.2   The motoneurons

There are two SP motoneurons per abdominal ganglion; their cell bodies are located on the ventral surface, medial to the exit point of the ventral nerve. Our knowledge of their mature morphology derives from single cell dye-fills in the adult (Taghert, unpublished results); our knowledge concerning their pattern of growth comes from TN-1 immunostaining in the embryo. Within the ganglion, their axons travel medially within the neuropil, meet along the midline, then exit the ganglion posteriorly via what will become the median nerve. In the abdominal ganglion, these events occur between embryonic stages 34% and 37%. By 39%, the growth cones of the two MNs are typically seen just approaching the strap (Fig. 4B). By 41%, they have almost without exception traveled onto the strap and each has bifurcated to grow laterally along it. By 50% they have reached the insertion point of the strap into the ectoderm and by 57% they have travelled across the bridge cells to their muscle target posterior to the spiracle.

2.2.3   The peripheral neuroendocrine cells

The two peripheral NE neurons that project into the TN are called L1 and L2. The L refers to link neurons; this nomenclature was initiated by Finlayson and Osborne (1969) in their previous studies of peripheral NEs in the stick insect *Carausius* which occur at link points between peripheral nerves. In *Manduca*, these neurons are found around the spiracle, in positions that are reasonably stereotyped (Fig. 6). We have first observed them at about 39%, when they lie close to each other on the posterior surface of the trachae that are growing upwards from the spiracle. By 42% (Fig. 4B), L1 has grown its axon to the bridge, by 45% to the strap insertion point, and by 47% it has reached the level of the ganglion. More details concerning its pattern of growth are described in a later section. L2 first projects an axon into the DN back to the ganglion where it projects processes into the neuropil and also up the ipsilateral connective to the next anterior ganglion. Around 52% it begins to send secondary axons, one peripherally towards the L1 cell body and one centrally, following the L1 axon along the TN.

2.2.4   The central neuroendocrine cells

The majority of neurons that project to the TN do so via the ipsilateral VN of

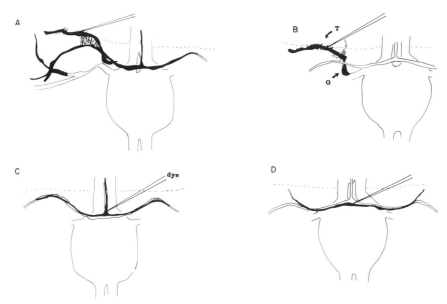

Fig. 6 Non-neural elements of the TN. Camera lucida drawings of Lucifer Yellow filled cells in living and prefixed preparations. (A) Dye-fill of the tracheoblast in living preparations reveals dye-coupling to cells of the strap and bridge on the same side of the animal. Note that dye has also spread across the midline and up the median nerve. (B) Morphology of the tracheoblast (T) is shown when dye-filling is performed following paraformaldehyde fixation. A specific strap cell (glioblast, G) is lightly coupled even following fixation. (C) The "motor" pathway of the nerve is revealed by dye-fills of strap cells in the anterior part of the nerve following fixation. (D) The "secretory" pathway is revealed when fills are made in the posterior part of the nerve following fixation. See text for further details.

the next anterior ganglion. These include three of the four bursicon-containing neurons and two (and later during metamorphosis six other) mid-line neurons. At all times while in transit to the TN, these axons are highlighted by TN-1 immunostaining (Fig. 4). Their growth cones can be seen first exiting the ganglion by about 36%; when they are halfway to the TN, they are consistently seen apposed a group of TN-1-positive cells (future glia?) that lie along a presumptive oblique external muscle. Passing this group of cells the growth cones continue travelling posteriorly across the segment boundary until they reach the TN. The site at which they join the developing nerve is also a constant feature; this is the insertion point of the strap into the ectoderm.

The growth cones (typically one leads two to three others) reach this point by 47–49% (Fig. 4D) and there undergo a morphological transformation. Whereas in transit the growth cones were small, thickened extensions of the axons, at the strap insertion point, they become considerably flattened, broad and display numerous far-reaching filopodia (Fig. 4E). Typically, they continue in this state for at least 8–9% of development before they finally resume growth. Each neuron now displays cell-specific directions of growth along the strap. The first growth cone grows laterally; the next two bifurcate and grow both centrally and laterally. The L1 cell also displays a striking number of filopodia in this region at this time but certain differences are noticeable (see below). This specific point of connection is of interest because this area of intense lamellopodial and filopodial activity is the site of what later becomes the thickened PVO. It will be composed of numerous neuronal ramifactions and glial cell projections.

3 Further details of TN differentiation

3.1 GLIA DEFINE FUNCTIONAL DOMAINS WITHIN THE TN

By the 50% stage of development, the TN neurons that will differentiate during embryogenesis have all reached the strap and begun to grow along it. The few strap cells that remain (circa 10) are situated in positions along the nerve which are not stereotyped. However, certain positions appear to be preferred; for example, three cells are seen near the "joint" between the TN and DN, and two cells are seen along the TN that lies dorsal to the ganglion. These observations suggest the possibility that the strap cells may be identifiable. Lucifer Yellow dye injections into living strap cells at this age demonstrate that they are all dye-coupled throughout the strap and bridge. (MW of the dye = 450.) Interestingly, this pattern of coupling is specific in that cells within the DN are not included.

If, prior to dye filling, embryos are first fixed—a treatment that often destroys dye-coupling (Thomas *et al.*, 1984)—a different pattern emerges. Two continuous pathways of strap cell processes are revealed (Fig. 6C and D). Fills of cells in the anterior half of the TN at the midline demonstrate a pathway that is identical to that taken by the MNs. That is the median nerve, the anterior half of the TN when dorsal to the ganglion, the posterior half of the TN as it proceeds towards the insertion point (Fig. 6C). Dye-coupling fades as the pathway turns towards the bridge (and the muscle target). Dye fills of strap cells in the posterior half of the TN at the midline highlight the pathway that is taken by all the secretory neurons. Dorsal to the ganglion, this pathway lies in the posterior half of the nerve; moving towards the insertion point, it switches to the anterior position. At the insertion point, the strap processes move off the nerve and insert into the nearby ectoderm (Fig. 6D).

Hence, these patterns of strap cell coupling correlate specifically with the motor and secretory domains of the TN (Fig. 7).

3.2   A SYNCYTIUM MAY BE INVOLVED IN TN MORPHOGENESIS

Just prior to the arrival of the bursicon cell, growth cones have reached the strap insertion point; a syncytium of three to four nuclei migrates into this position from a lateral location near the spiracle. Its final position is roughly constant, anywhere up to approximately $25\,\mu$ anterior to the strap insertion point. We believe this variability may be a dissection artifact. Ramifications of the L1 axon (see below) and perhaps also filopodia from the bursicon cell growth cones show a strong tendency to adhere to this cell (Fig. 9A and B) and this tendency may last well into late embryonic stages. Post-embryonically this syncytium can be seen at a position near the TN just near its attachment point with the bursicon axons from the VN. This tendency to adhere is seen in the behaviour of the neuronal filopodia which show a selective preference for growing along the surface of the syncytium to the exclusion of most other surfaces and also, in a tendency of the growth cones or lamellopodial extentions of the axons to appear in close apposition to it. In Section 4.1 we document selective filopodial adhesion for cell L1.

Interestingly, this syncytium becomes dye-coupled to cells in the strap (Fig. 6A). This pattern is observed in living embryos and consists of strap cells in both the motor and secretory "domains". In order to visualize the shape of the tracheoblast, it is first necessary to pre-fix the embryo. Even then, slight dye coupling occurs with one identifiable strap cell (Fig. 6B) which is closely apposed to the tracheoblast cell at its leading medial edge. The tracheoblast syncytium therefore appears to be involved in TN morphogenesis by virtue of its specific association with the L1 cell and by virtue of its dye-coupling to strap cells. Further elucidation of the exact role played by this syncytium must await its experimental ablation.

4   Cell differentiation of L1: an identified peptidergic neuron

In the preceding sections we have described in both general and specific terms the construction of the TN in the abdominal segments of *Manduca*. Motoneuron and neuroendocrine neuron axons arrive and grow into a previously elaborated framework of cells to create a stereotypically formed nerve. Here we consider the differentiation of a single member of the neural complement in order to more carefully examine the details of the developmental schedule that is employed. We would like to know at a cellular and, ultimately, at a

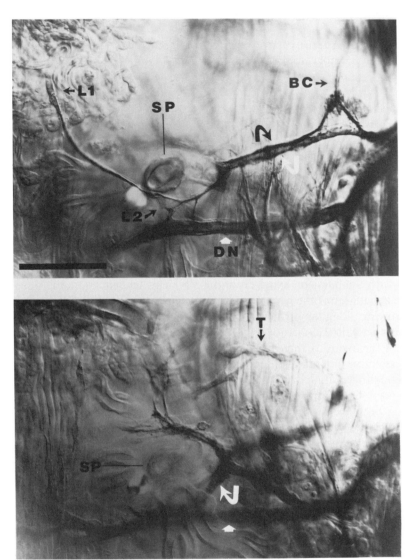

Fig. 7   Two focal planes of a TN-1-stained embryo at 62% of development. The field
of view is to the left of abdominal ganglion 3. (*Upper*) The TN is visible from the L1
cell body (L1) to the ganglion at the far right. The L1 axon loops posteriorly around
the level of the spiracle (SP) where it is still in contact with cell body of the peripheral
NE cell L2. The motor and secretory pathways within the nerve are clearly established
and are highlighted by a dark curved arrow (secretory) and a light curved arrow
(motor). The axons of the bursicon cells (BC) can be seen anastomosing to grow both
centrally and laterally along the secretory tract. DN: dorsal nerve. (*Lower*) A more
ventral focal plane of the same field. The syncytial cell (T) is seen close to the segment
boundary, and the approach of the motor axons to their muscle target is now in focus
(light curved arrow). (Scale bar: 50 μ.)

molecular level, how single neurons come to assume mature and unique cellular phenotypes. We have focussed on neuron L1, a peripheral neuroendocrine neuron. This choice was made for two specific reasons. Firstly, a
monoclonal antibody raised against the small cardioactive peptide$_B$ of
Aplysia (Lloyd, 1986) specifically stains this and other suspected CAP
neurons in *Manduca* (Copenhaver and Taghert, unpublished results) and
hence, it is a marker for biochemical differentiation. Secondly, the configuration of this neuron in the living animal is such that it gives the experimenter
many of the advantages of a neuron in tissue culture. In other words its processes are in a quasi-two-dimensional array and that, rather than being partially or completely "hidden" within the neuropil, all L1 processes are
peripheral and hence accessible to the experimeter.

4.1 MORPHOLOGICAL DIFFERENTIATION OF L1

A timeline of morphological growth for neuron L1 is shown in Figure 8.
These are camera lucida drawings of the neuron following Lucifer Yellow
dye injection via a microelectrode. L1 (and L2, another peripheral NS cell) is
first identifiable at about 39%, in a position just posterior to the developing
trachae that grow out from the spiracle. The initial contiguity of the L1 and
L2 cell bodies suggests the possibility that they may derive from a single precursor cell in the underlying ectoderm. With time, the L1 cell body moves
laterally and anteriorly to a position closer to the dorsally-placed heart of the
animal. The extent of lateral migration is segment-specific. In segments T1
through to A1, the L1 cell body remains close to its initial position; in segments A2 through to A7, the cell body is approximately halfway between the
spiracle and the heart (Fig. 7A). In all cases, it projects a growth cone that
grows centrally around the posterior side of the trachae and towards the
bridge cells, which it reaches by 42%.

Simultaneously, L1 grows one or more distal growth cones in the direction
of the developing heart; the number of distal processes is again a function of
segment. In T1–A1, only one distal process is elaborated; in the more posterior segments, two are usually seen. In the mature state, the distal axons are
draped along trachae and form neurohaemal structures close to or within the
heart. By 45%, the centrally-directed growth cone has reached the strap
insertion point and is following the strap cells centrally towards the ganglion.
By about 47%, it has reached the level of the ganglion and typically pauses at
this site for 8–12% of development.

The L1 growth cones does not become enlarged or lammelopodial, and
does not give any other indication of pausing as if at a choice point (Bastiani
et al., 1984). Interestingly, these events are occurring in cell L1 at this time,

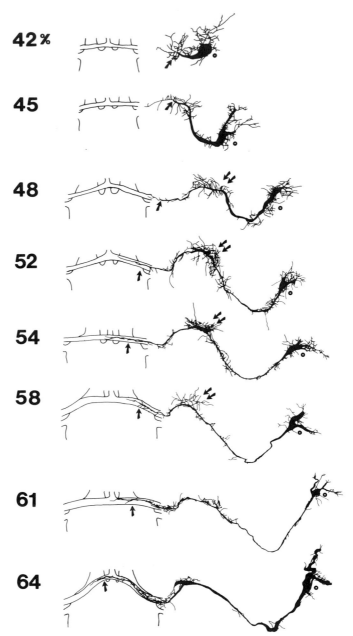

42%

45

48

52

54

58

61

64

Fig. 8   Morphogenesis of the L1 cell in abdominal segments 2 or 3. Camera lucida drawings of Lucifer Yellow-filled cells at the indicated percentages of development. Asterisks marks the position of the cell body; single arrow marks the furthest extent of the leading growth cone; double arrows mark the position of the syncytium (not seen) and strap insertion point. Note the large filopodial spray that is often seen at this location. (Scale bar: 50 μ.)

but they happen away from its growth cone, back at the strap insertion point (Figs 8 and 9). This insertion position is not a major choice point for L1 branching and the morphological specialization at that site therefore presumably subserves a different function. In abdominal segments 2–7, L1 growth resumes across the midline in the TN between 55–60%. By a later stage (circa 90%), L1 has reached the strap insertion point on the contralateral side of the midline and has terminated growth. The filopodial and lammelopodial activity of the L1 axon reflects its preference to adhere to the syncytial cell (Fig. 9A and B). This occurs despite the fact that the syncytium appears long after the L1 growth cone has passed through the strap insertion point. In contrast, the MNs, which have passed through this point simultaneously with L1 but growing in the opposite direction, do not exhibit this behaviour. It should be noted that this is the site of the future nerve thickening that is termed the perivisceral organ (PVO) in other insect orders (Raabe et al., 1974).

L1 shows affinity for a second syncytium; like the first, this syncytium lies close to the anterior segment boundary but its position is within the dorsal aspect of the segment (Fig. 9B). Given this configuration, the distal processes of the L1 cell appear to adhere closely to its surface as they grow towards the heart. Neither syncytium appears to be the final target of the L1 cell; rather they are structures that L1 interacts with along its direction of growth. The nature of the interaction is at present hypothetical, but may include guidance, i.e. as a landmark cell (Bastiani et al., 1985), provision of rigid structural support, or a more loosely definable trophic support.

The mature L1 neuron displays numerous varicosities and processes grown in parallel along the length of TN and adjoining nerves (Fig. 9C). These structures represent the points of transmitter (hormone) release— neurohaemal sites. These mature morphological phenotypes are produced between 60 and 80% of development, before the L1 neuron has completed its limit of total axon growth. Within the TN, the L1 pathway follows the secretory "domain" that was described in dye-fills of strap cells (section 3.1). The developmental queston as to which cell type (neuronal or glial) first defines such pathways for the other must await experimental manipulation.

4.2   BIOCHEMICAL DIFFERENTIATION OF L1

SCP-like immunoreactivity is first expressed in L1 at about 42% along its axon and cell body (Wall and Taghert, unpublished results). It is the first neuron in the animal to do so. The first segments that show expression are T2–A1; within a short time, all others are similarly expressive (Fig. 9C). By comparison with the onset of 5-HT expression (Taghert and Goodman,

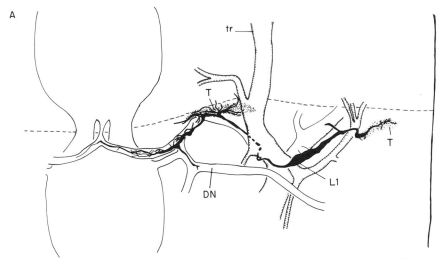

Fig. 9   Details of L1 differentiation. (A) Camera lucida drawing of the L1 cell in segment A1 at 59% as it lies draped over trachea (tr) that lie to the side of the ganglion.
The growth cone has reached the midline and is starting to project a secondary process into the dorsal nerve (DN). Note the profusion of filopodia over the closely posi-
tioned syncytium (T) that is shaded by dots. A second syncytium is contacted by the
distal (to the right) L1 process.

1984) and proctolin expression (Keshishian and O'Shea, 1985) in insect neurons, this is a rapid developmental onset of expression. It follows axonogene-
sis by only 2–3% (as compared to approximately 15% in the studies cited
above). The possibility that the expression of this antigen by diverse neurons
is strictly a function of age must await determination of neuronal birthdates.
As yet we know only that L1 is born by at least 39%. Neurons that innervate
the midgut of *Manduca*, also born in the mid-30s%, do not begin expressing
this antigen until approximately 65% (Copenhaver and Taghert, unpublished results). Thus, it appears that the schedule of biochemical differentiation for neurons which express a similar peptide antigen differs according to
the cell lineage or the cell position (and therefore the cell interactions) that
different cells may have.

   The basis and/or reasons for the early expression by cell L1 may become
clearer through experimental manipulation (Wall and Taghert, unpublished
results). One interesting possibility for the function of "early" transmitter
expression is suggested by recent work on molluscan neuronal growth cones
in culture (Haydon *et al.*, 1984). Specific growth cones appear to be sensitive
to certain neurotransmitters (e.g. 5-HT and dopamine) and to cease growth
in their presence. These authors speculate that NTs may have developmental

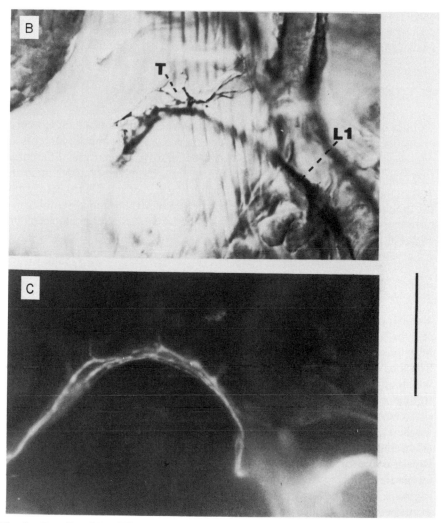

Fig. 9   Details of L1 differentiation. (B) Photomicrograph of the preparation that was drawn in (A). Note the filopodial spray of the L1 cell upon the syncytium (T). (C) The L1 cell stained with the anti-SCP$_B$ Mab at 90%. Note the presence of numerous parallel processes that run within the nerve and that are varicose. Note also the lack of substantial processes where the L1 filopodial spray previously occurred. (Scale bar: 60 μ.)

functions, including influencing axon guidance and specific connectivity through their ability to affect growth cone navigation. Perhaps the relatively early embryonic expression of SCP-like material in L1 serves a similar function with regard to the arrival of the bursicon cells at the point of the L1 filopodial spray.

5   Inferences and conclusions

In this chapter we described the embryonic development of a simple neuro-endocrine effector organ, the transverse nerve, in abdominal segments of *Manduca*. These results suggest a pattern of stereotyped neuronal differentiation that includes specific cell interactions between developing NE neurons and glial precursors. In addition, numerous features of the TN trajectory and position appear to be "prefigured" by the prior arrival and assemblage of non-neuronal cells into cellular pathways that are later taken by the growing axons of the TN neurons.

5.1   TN POSITION AND TRAJECTORY ARE PREFIGURED BY NON-NEURONAL CELLS

The hypothesis that axonal pathways may be laid down by non-neuronal cells before the arrival of neuronal growth cones is one of many that have been posited to explain stereotyped nerve formation during development. Specifically, Singer *et al.* (1979) proposed the "blueprint hypothesis" in which non-neural neuroepithelial cells provide a highway for later differentiating neurons to use as axonal pathways. In the past 6 years, studies on insect embryos (Goodman and Bate, 1981; Bentley and Keshishian, 1982; Taghert *et al.*, 1982; Bastiani *et al.*, 1985) and more recently in fish embryos (Kuwada, 1986; Myers *et al.*, 1986) have provided insights into the mechanisms of nerve formation and growth cone guidance *in vivo*. These studies have emphasized the importance of cell surface interactions between the developing neuron and the local cellular environment. Sources of guidance that have been specifically implicated include single neurons, neuroepithelial cells (presumptive glia) (Bastiani, Doe and Goodman, in press), muscle pioneers (Ball *et al.*, 1985) and, in addition, elements of the overlying basement membrane and underlying epithelium. These later elements are thought to be in the form of gradients of adhesivity (Nardi, 1983; Berlot and Goodman, 1984; Caudy and Bentley, 1986) whereby neurons may obtain polarity cues.

What may be novel about the formation of axon pathways in the TN of *Manduca* is the presence not of discontinuous, single landmark cells, or the use of a two-dimensional sheet of cells, but the assemblage of a *continuous* pathway composed of *non-neuronal* cells (the strap and the bridge) that prefigure the exact position and trajectory of the future TN. An alternative explanation is that within the collection of strap cells are identified landmarks which are used by specific neurons. As yet, we do not know if the cells of the strap are homogeneous or whether they have already differentiated into distinct cell types before the neurons arrive. Such distinctions could be used to guide separate neuronal growth cones onto separate pathways within the developing nerve.

Ho and Goodman (1982) demonstrated that in grasshopper embryos, the transverse nerve is pioneered by the SP MNs. In moth embryos, we have found that the "motor" pathway is likewise pioneered by these same MNs, but that the L1 cells are simultaneously pioneering a separate pathway from the periphery inwards. Moreover, in grasshoppers, the substrate used by the MNs appears to be a single muscle pioneer cell, around which undifferentiated mesoderm collects to form a transverse muscle anlage. No such muscle develops in moth embryos and we have not found a single cell that appears comparable to the grasshopper muscle pioneer. Further characterization of the strap will allow a more comprehensive comparison.

5.2   FORMATION OF PERIPHERAL NERVE ANASTOMOSES

Peripheral nerves in *Manduca* are extensively cross-connected via anastomoses; these are used by individual neurons to navigate from one nerve pathway to another. These connections are stereotyped in number and position. They arise by at least two different mechanisms: (1) The bridge serves as an interchange where numerous anastmoses are formed. These glial cells prefigure future connections, much as the strap prefigures TN position and trajectory; (2) The anastomosis between the TN and VN (the PVO), which is used by the bursicon cells and others, appears to be marked at a certain location in the ectoderm—this is the strap insertion site. We tentatively interpret situation (2) to indicate "rendevousing" behaviour on the part of the neuronal growth cones, as opposed to the utilization of preformed connections.

5.3   STEREOTYPED DEVELOPMENT OF NEUROENDOCRINE CELLS

Analysis of the development of the L1 neuron suggests that neuroendocrine neurons differentiate in a manner that is very similar to that previously documented for insect motoneurons and interneurons (Raper *et al.*, 1983; Ho and Goodman, 1982; Taghert and Goodman, 1984). That is to say, morphological development proceeds smoothly without evidence of error in guidance, and with suggestions of specific cell-surface interactions between the growth cone and elements of the cellular environment that are encountered along the way. The biochemical differentiation of L1 (as indicated by the onset of its putative transmitter, a cardioactive peptide-like substance) occurs after the initiation but before the completion of axon outgrowth. This schedule is similar to that seen in serotonergic neurons in grasshopper embryos (Taghert and Goodman, 1984), other peptidergic neurons in *Manduca* (Copenhaver and Taghert, unpublished results), and in other animals (e.g. Pickel *et al.*, 1982). Particularly attractive features of the L1 neuron for a mechanistic analysis of transmitter differentiation include not only its stereotyped schedule, but also the complete accessibility of its processes for experimental manipulation. Such experiments are currently in progress in our laboratory.

5.4   DO NEUROENDOCRINE CELLS HAVE TARGETS?

Developing neurons are dependent on target cells for a variety of developmental choices. Depending on the specific example studied, these choices include survival (Hamburger, 1975), guidance and arrest of growth (Ball *et al.*, 1985), and the exact positioning of the terminal arborization (Murphy and Lemere, 1984). In the case of neuroendocrine neurons that lack closely apposed target cells, what are the mechanisms that regulate their viability, arrival at the future neurohaemal site, and the positioning and number of their hormone release sites?

During the development of the pituitary gland in vertebrates (analogous to the TN in having a neurohaemal function), non-neural cells termed pituicytes appear to play a pivotal role in the differentiation of neurons (Dellman *et al.*, 1981). Before the neuronal growth cones appear from the hypothalamus, the pituicytes are present in a quiescent mode. After the arrival of the neurons, these pituicytes switch to a secretory mode. Dellman *et al.* (1981) interpret this to mean that these non-neuronal cells are effecting the outgrowth or differentiation of the neurons. In the case of the developing TN, we have observed events that suggest specific interactions between the growing axons of NE cells and non-neural cells that lie along and near the strap. These observations of normal development imply a flow of information from strap cells to neurons. We hypothesize that some or all of these non-neural cells may substitute as fictive targets for the developing NE cells that must produce blind, non-synaptic varicose endings within the nerve.

For the purpose of designing specific experiments, we hypothesize three forms of information that may be utilized and which may be tested.

(1) *Axon guidance information.* Within the assemblage of strap cells, there

probably exists directional information that is utilized by the developing neurons as each takes cell-specific pathways along and within the strap.

(2) *Induction of terminal growth*. Here we posit the existence within the strap of signals to direct the growth cones to stop growing and to induce them to produce secondary and tertiary branches that carry terminal varicosities. Stop signals for insect motoneurons have been localized previously in single cells (Ball *et al.*, 1985). The signals that induce varicosity formation are unknown.

(3) *Regulation of neurotransmitter expression*. Transmitter expression has been shown repeatedly to be a plastic phenotype (Black, 1978; Patterson, 1978); in invertebrates, specific evidence is lacking. In the case of the TN neurons, L1 shows a normal onset of transmitter expression as it enters the strap and begins to interact with specific cells. As mentioned previously, analysis of the L1 cell *in vivo* may allow us to explore the range of transmitter regulation directly.

5.5   PROSPECTS FOR A MOLECULAR ANALYSIS OF TN FORMATION

We have described the development of a peripheral nerve by analysing the behaviour of individual neurons and non-neural cells. A recurring theme in these observations is the remarkably precise recognition that individual embryonic cells display as they explore their environment during their differentiation. For example, the MN and NE cells navigate smoothly along the cells of the bridge and strap, but within different domains. Furthermore, these NE neurons later produce varicose endings at non-random positions along the nerve (Wall and Taghert, unpublished results). These precise cellular events imply equally precise underlying molecular recognition, the study of which would greatly expand our understanding of neuronal differentiation. It must be conceded, however, that these molecules, expressed by small numbers of neurons or non-neural cells, are likely to be extremely rare. One possible technical avenue of approach would be the use of monoclonal antibodies.

The TN-1 Mab is an extremely useful histological stain in *Manduca* in that it specifically and strongly labels subsets of neurons and non-neuronal cells in developing embryos. TN-1 is a cell surface antigen as indicated by its binding to antibody in living embryos (Carr and Taghert, unpublished results) and may be related to a family of cell-surface glycoproteins recently discovered in grasshopper embryos (Goodman, personal communication). TN-1 has proved invaluable in this analysis of TN formation (Carr and Taghert, in preparation) and as well in describing the embryonic formation of the stomatogastric nervous system in *Manduca* embryos (Copenhaver and

Taghert, in preparation). Other Mabs from this fusion hold equal promise in further elucidating the cellular interactions that underlie the formation of this nerve. At least four other Mabs also show some cell-specific staining of neuronal and non-neural elements of the developing TN. For example, Mab TN-4 specifically stains three strap cells near the strap insertion point. Some of the Mabs, which are currently helping to elucidate cellular events, are also indicating molecules that may mediate some of these processes. This prospect will be addressed in the near future.

5.6   THE DEVELOPMENT OF A NEUROHAEMAL ORGAN

For many years, neurohaemal organs have been necessary and interesting neural structures for study in the context of the endocrine function of the nervous system. Consideration of neurohaemal organs as neuronal targets, however, presents them in a different light. As regards neuronal development, they become equivalent to a muscle, gland or nerve cell through their innervation by specific neurons. We have seen that a prospective neurohaemal organ has a form prior to the arrival of any neural element and that further differentiation of the neurons that project to this organ is as stereotyped as that described for more orthodox neural elements. Thus, within this simple, anatomically distinct organ, the transverse nerve, there occur many of the basic features of neuronal development: cell-specific axon guidance, the halting of axon growth, the creation of secondary and tertiary branching patterns and the induction of release elements (varicosities) along axons. What makes a neurohaemal organ attractive as a simple system in which to study these processes is that it represents a coherent and relatively homogeneous tissue, one that is not contaminated by cells performing other functions (e.g. muscles, other neuron types, epithelium, etc.). Future studies will utilize this numerical and functional simplicity to further analyse the cellular and molecular events that underlie neuronal differentiation.

Acknowledgements

We thank Steven Kemp, Boris Masinovsky and Dennis Willows for the gift of the anti-SCP Mab. Philip Copenhaver helped perform some of the first descriptions of the L1 neuron and has also contributed useful suggestions and criticisms. We thank our colleagues for sharing unpublished results. We gratefully acknowledge support of the studies described in this chapter by an NSF Pre-Doctoral Grant (JNC), and by funds from The McDonnell Center for Cellular and Molecular Neurobiology (Washington University School of Medicine) and by a grant from the NIH No. NS21749 to P.H.T. who is a Sloan Fellow.

References

Ball, E., Ho, R. K. and Goodman, C. S. (1985). Development of neuromuscular specificity in the grasshopper embryo: guidance of motoneuron growth cones by muscle pioneers. *J. Neurosci.* **5,** 1808–1819.

Bastiani, M. J., Raper, J. A and Goodman, C. S. (1984). Pathfinding by neuronal growth cones in grasshopper embryos. III. Selective affinity of the G growth cone for the P cells within the A/P fascicle. *J. Neurosci.* **4,** 2311–2322.

Bastiani, M. J., Doe, C. Q., Helfand, S. L. and Goodman, C. S. (1985). Neuronal specificity and growth cone guidance in grasshopper and *Drosophila* embryos. *Trends Neurosci.* **8,** 257–266.

Bate, C. M. (1976). Embryogenesis of an insect nervous system. I. A map of the thoracic and abdominal neuroblasts in *Locusta migratori. J. Embryol exp. Morphol.* **35,** 107–123.

Bate, C. M. and Grunwald, E. (1981). Embryogenesis of an insect nervous system. II. A second class of neuronal precursors and the origin of the intersegmental connectives. *J. Embryol exp. Morphol.* **61,** 317–330.

Bentley, D., Keshishian, H., Shankland, M. and Torian-Raymond, A. (1979). Quantitative staging of development of embryonic development of the grasshopper. *J. Embryol exp. Morphol.* **54,** 47–74.

Bentley, D. and Keshishian, H. (1982). Pathfinding by peripheral pioneer neurons in grasshoppers. *Science* **218,** 1082–1088.

Berlot, J. and Goodman, C. S. (1984). Guidance of peripheral neurons in the grasshopper: an adhesive hierarchy of epithelial and neuronal surfaces. *Science* **223,** 493–496.

Black, I. B. (1978). Regulation of autonomic development. *Ann. Rev. Neurosci.* **1,** 183–214.

Blair, S. (1982). Blastomere ablation and the developmental origin of identified monoamine-containing neurons in the leech. *Dev. Biol.* **95,** 65–78.

Carpenter, E. M. and Hollyday, M. (1986). Defective innervation of chick limbs in the absence of presumptive Schwann cells. *Soc. Neurosci.* **12,** 1210.

Carrow, G. M., Calabrese, R. L. and Williams, C. M. (1984). Architecture and physiology of cerebral neurosecretory cells. *J. Neurosci.* **4,** 1034–1044.

Caudy, M. and Bentley, D. (1986). Pioneer growth cone morphologies reveal proximal increases in substrate affinity within leg segments of grasshopper embryos. *J. Neurosci.* **6,** 364–379.

Copenhaver, P. F. and Truman, J. W. (1986). Metamorphosis of the cerebral neuroendocrine system in the moth *Manduca sexta. J. comp. Neurol.* **249,** 186–204.

Dellman, H. C., Slkora, K. and Castell, M. (1981). Fine structure of the rat supraoptic nucleus and neural lobe during pre- and post-natal development. *In* "Neurosecretion: molecules, cells and systems" (Eds D. Farner and K. Lederis), pp. 177–186. Plenum Press, New York.

Doe, C. Q. and Goodman, C. S. (1985). Early events in insect neurogenesis. I. Development and segmental differences in the pattern of neuronal precursor cells. *Dev. Biol.* **111,** 193–205.

Finlayson, L. and Osborne, M. P. (1969). Peripheral neurosecretory cells in the stick insect (*Carausius morosus*) and the blow-fly (*Phormia terranovae*). *J. Insect Physiol.* **14**, 1793 1811.

Goodman, C. S. and Bate, C. M. (1981). Neuronal development in the grasshopper. *Trends Neurosci.* **4**, 163–169.

Hamburger, V. (1975). Cell death in the development of the lateral motor column in the chick embryo. *J. comp. Neurol.* **160**, 535–546.

Haydon, P. G., McCobb, D. P. and Kater, S. B. (1984). Serotonin selectively inhibits growth cone motility and synaptogenesis of specific identified neurons. *Science* **226**, 561–564.

Ho, R. K. and Goodman, C. S. (1982). Peripheral pathways are pioneered by an array of central and peripheral neurones in grasshopper embryos. *Nature* **297**, 404–407.

Keshishian, H. and O'Shea, M. (1985). The acquisition and expression of a peptidergic phenotype in the grasshopper embryo. *J. Neurosci.* **5**, 1005–1015.

Kuwada, J. (1986). Cell recognition by neuronal growth cones in a simple vertebrate embryo. *Science* **233**, 740–746.

Lewis, G. W., Miller, P. L. and Mills, P. S. (1973). Neuromuscular mechanisms of abdominal pumping in the locust. *J. exp. Biol.* **59**, 149–168.

Lloyd, P. E. (1986). The small cardioactive peptides: a class of modulatory neuropeptides in *Aplysia*. *Trends Neurosci.* **9**, 428–431.

Murphey, R. K. and Lemere, C. A. (1984). Competition controls the growth of an identified axonal arborization. *Science* **224**, 1352–1355.

Myers, P. Z., Eisen, J. S. and Westerfield, M. (1986). Development and axonal outgrowth of identified motorneurons in the zebrafish. *J. Neurosci.* **6**, 2278–2289.

Nardi, J. (1983). Neuronal pathfinding in developing wings of the moth *Manduca sexta*. *Dev. Biol.* **95**, 163–174.

Oland, L. A. and Tolbert, L. P. (1986). Reduction of glial population by gamma-irradiation disrupts development of glomeruli in *Manduca sexta* antennal lobe. *Soc. Neurosci.* **12**, 929.

Patterson, P. H. (1978). Environmental determination of autonomic neurotransmitter functions. *Ann. Rev. Neurosci.* **1**, 1–17.

Pickel, V. M., Sumal, K. K. and Miller, R. J. (1982). Early pre-natal development of Substance P- and enkephalin-containing neurons in the rat. *J. comp. Neurol.* **210**, 411–431.

Raabe, M. (1982). "Insect neurohormones". Plenum Press, New York.

Raabe, M., Baudry, N., Grillot, J. P. and Provensal, A. (1974). The perisympathetic organs of insects. *In* "Neurosecretion: the final common pathway" (Eds F. Knowles and L. Vollrath) pp. 60–71. Springer Verlag, New York.

Raper, J. A., Bastiani, M. J. and Goodman, C. S. (1983). Pathfinding by neuronal growth cones in grasshopper embryos. I. Divergent choices made by growth cones of sibling neurons. *J. Neurosci.* **3**, 20–30.

Singer, M., Nordlander, R. H. and Egar, M. (1979). Axonal guidance during embryogenesis and regeneration in the spinal cord of the newt. "The Blueprint Hypothesis" of neuronal pathway patterning. *J. comp. Neurol.* **185**, 1–22.

Sulston, J. E. and Horwitz, H. R. (1977). Post-embryonic lineages of the nematode, *Caenorhabditis elegans*. *Dev. Biol.* **56**, 110–156.

Taghert, P. H. and Truman, J. W. (1980). Morphology and function of neurons that differentiate *de novo* during adult development in the tobacco hornworm, *Manduca sexta*. *Soc. Neurosci.* **7**, 743.

Taghert, P. H. (1981). Identification of specific peptidergic neurons in the moth, *Manduca sexta*. Ph.D. Thesis, University of Washington.

Taghert, P. H. and Truman, J. W. (1982). Identification of the bursicon-containing neurons in abdominal ganglia of the tobacco hornworm, *Manduca sexta. J. exp. Biol.* **98**, 385–401.

Taghert, P. H., Bastiania, M. J., Ho, R. K. and Goodman, C. S. (1982). Guidance of pioneer growth cones: filopodial contacts and coupling revealed with an antibody to Lucifer Yellow. *Dev. Biol.* **94**, 391–399.

Taghert, P. H., Tublitz, N. J., Truman, J. W. and Goodman, C. S. (1983). Generation of monoclonal antibodies to the neurohaemal transverse nerve of the moth, *Manduca sexta. Soc. Neurosci.* **9**, 256.

Taghert, P. H. and Goodman, C. S. (1984). Cell determination and differentiation of identified serotonin-immunoreactive neurons in the grasshopper embryo. *J. Neurosci.* **4**, 989–1000.

Taghert, P. H.. Carr, J. N., Wall, J. B. and Copenhaver, P. F. (1986). Embryonic formation of a simple neurosecretory nerve in the moth. *In* "Third International Conference of Insect Neurochemistry and Neurophysiology". (Ed. A. Borkovic). Humana Press, New York (in press).

Taylor, H. and Truman, J. W. (1974). Metamorphosis of the abdominal ganglion in the tobacco hornworn, *Manduca sexta. J. comp. Physiol.* **90**, 367–388.

Thomas, J. B., Bastiani, M. J., Bate, C. M. and Goodman, C. S. (1984). From grasshopper to *Drosophila*: a common plan of neuronal development. *Nature* **310**, 203–207.

Truman, J. W. (1973). Physiology of insect ccdysis. III. Relationship between the hormonal control of eclosion and tanning in the tobacco hornworm, *Manduca sexta. J. exp. Biol.* **57**, 805–816.

Tublitz, N. J. and Truman, J. W. (1985). Insect cardioactive peptides. I. Distribution and molecular characterization of two cardioacceleratory peptides (CAP) in the tobacco hawkmoth, *Manduca sexta. J. exp. Biol.* **114**, 365–379.

Tyrer, N. M. (1970). Quantitative estimation of the stage of embryonic development in the locust, *Schistocerca gregaria. J. embryol exp. Morphol.* **23**, 705–718.

Thermoregulation and Heat Exchange

Timothy M. Casey

Department of Entomology and Economic Zoology, New Jersey Agricultural Experiment Station, Cook College, Rutgers University, New Brunswick, New Jersey 08903, USA

1   Introduction

Temperature has a profound influence on virtually all physiological systems. The small size of insects causes rapid rates of heat exchange with the environment while at the same time allowing them to utilize thermal microenvironments not available to larger animals (Willmer, 1982). Consequently, the thermal balance of insects is of interest from a variety of biological viewpoints. Although insects have long been known to generate considerable temperature excesses as a byproduct of their metabolism (see Heinrich, 1981a for a historical overview), it wasn't until relatively recently that Heath and Adams (1965) first demonstrated that sphinx moths regulated their thoracic temperature during flight. From Heinrich's (1974a) first review of the subject, aspects of insect thermoregulation have been reviewed repeatedly (Kammer and Heinrich, 1978; May, 1979, 1983; Willmer, 1982). *Insect Thermoregulation* (edited by B. Heinrich, based on a symposium held in 1980) attempted to examine the phenomenon from all perspectives. Since that time the scope of

ADVANCES IN INSECT PHYSIOLOGY VOL. 20
ISBN 0–12–024220–6

questions, methods of analysis and range of implications of the results for understanding insect biology continue to grow at an ever increasing pace. I will not attempt to synthesize the area of insect thermoregulation in this short review, but rather try to update it, based on the large number of studies which have appeared over the last few years. I will focus on information from biophysical, physiological and behavioural perspectives, and to try to indicate where integration of these approaches has increased our understanding of thermal biology.

2 Avenues of heat exchange

If an insect thermoregulates, it must control the temperature of its body relative to a varying environment by regulating rates of heat flow to and from its surroundings. The rate of heat gain must balance the rate of heat loss to achieve a steady state; otherwise body temperature will rise or fall based on the magnitude of heat gain or loss. However, it is often difficult to evaluate avenues of heat exchange because (a) the characterization of the environment is too simple, (b) the actual thermal properties of the animal are only approximately known, (c) the dynamics of heat flow in the body (physiological mechanisms of thermoregulation) are not known, or (d) the heat exchange model being employed is too simple. Any or all of these factors may confound analysis of heat exchange.

Body temperature is a complex function of several physical factors, and can be described by the equation:

$$T_b = T_a + / - R + / - C + / - G + M - E$$

where R = radiation, C = convection, G = conduction, M = metabolism and E = evaporation (Fig. 1). Heat exchange by radiation, conduction or convection can be either positive or negative depending on the environmental conditions, while metabolism is always positive (although often insignificant) and evaporation is always negative.

Animals are normally characterized as ectothermic or endothermic, depending upon whether the heat utilized for thermoregulation is obtained from external sources or generated internally, through metabolism. Ectotherms usually regulate T_b behaviourally, whereas endotherms regulate body temperature physiologically. In the insects, endothermy is common but it is not obligatory and is always periodic. Moreover, due to the high levels of heat production needed to raise body temperature, insects capable of sustaining high body temperatures endothermically routinely supplement metabolic heat production by behavioural thermoregulation. There are obvious cases

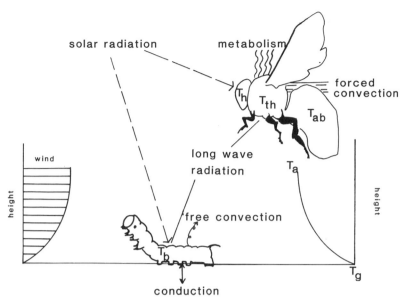

Fig. 1   Schematic diagrams for the avenues of heat exchange. (*a*) An ectothermic insect on the ground. (*b*) An endothermic insect during flight.

of strict ectothermy and strict endothermy among the insects but in many cases the characterization falls somewhere in between.

As a result of detailed laboratory studies, several generalizations about heat exchange in insects are possible. Radiative heat gain and convective heat loss represent the major avenues of heat exchange, except in insects which are conspicuously endothermic. Except in special cases, evaporative cooling is not a major mechanism for controlling body temperature (Seymour, 1974). In sunshine, the temperature excess $(T_b - T_a)$ is directly related to radiant intensity (Parry, 1951; Digby, 1955). Various factors including size, shape, surface area, coloration, and degree of pubescence will affect the capacity to absorb radiant heat. Convective heat exchange will vary with size, shape, orientation and surface properties of the insect, as will the characteristics of the air, including velocity and degree of turbulence. At high wind velocities (forced convection), the convection heat exchange is proportional to the square root of the velocity. At low wind velocity (0–30 cm/s) free convection effects predominate and convective heat transfer is almost independent of velocity (Digby, 1955; Church, 1960).

Physiologists have used Newton's law of cooling to quantify the rates of heat exchange in insects. This relationship describes the rate of temperature change of an object as:

$$\mathrm{d}T/\mathrm{d}T = \mathrm{K}\ (T_{th} - T_a)$$

where K is the Newtonian cooling constant and $T_{th} - T_a$ is the temperature difference between the thorax and the environment representing the driving force for temperature change. The cooling constant is determined empirically by measuring the cooling curve of an animal in the laboratory under controlled environmental conditions (Bakken, 1976a, b). Multiplying the cooling constant by specific heat yields thermal conductance (C), the rate of heat flow per degree of driving force. The equation now becomes:

$$\mathrm{d}H/\mathrm{d}T = C\,(T_b - T_a).$$

This relationship indicates that heat loss is a linear function of a constant and a temperature difference. A large body of data is now available from cooling experiments with insects from a variety of taxa (May, 1976b; Bartholomew, 1981). In general, rates of passive cooling are strongly inversely related to body mass, due to the increase in surface to volume ratio as size decreases and are generally similar for insects from different taxa at any given mass. Insulation on the thorax of large endothermic insects significantly reduces thoracic conductance (Church, 1960; Heinrich, 1971a).

The Newtonian cooling model is an oversimplification of the process of heat exchange because neither the conductance nor the temperature difference is well defined in the natural environment; thus, the model provides little insight into heat transfer (Tracy, 1972). Conductance will vary with the characteristics of the environment (i.e. wind speed, which affects the convection coefficient) and, similarly, the environmental temperature T_e is determined both by the animal and by the environment (Bakken and Gates, 1975). However, since radiative, conductive and convective heat exchange are all approximately linearly related to a temperature differential (Bakken and Gates, 1975), utilizing this model has been useful, particularly in the absence of significant radiant heat (May, 1983).

The environmental temperature surrounding an insect is complex and difficult to evaluate because the different avenues of heat exchange utilize different temperature gradients (e.g. radiant temperature, air temperature, substrate temperature; Fig. 1). In addition, temperature and wind velocity are difficult to measure accurately in the field because they vary considerably with distance from a substrate (Fig. 1). The actual temperature and wind conditions also depend on the characteristics of the substrate (size, shape, surface texture) and may be modified by the insect itself. Thus, the body of an insect may be exposed to a gradient of wind and temperature depending on its position in the environmental boundary layer. These factors have limited the use of a quantitative approach to evaluation of the thermal environment, and studies of insect thermoregulation have routinely reported body temperature of insects as a function of shade air temperature (see reviews by Heinrich, 1974a; Casey, 1981a; Kammer, 1981; May, 1983). Although these data can demonstrate whether or not thermoregulation occurs, they give little insight into the actual mechanisms of heat exchange.

While biophysical approaches have long been used in the laboratory (Parry, 1951; Digby, 1955; Church, 1960), the complexity and number of assumptions associated with quantifying heat balance relationships have, until recently, limited their utility for field studies of insect thermoregulation. However, studies utilizing biophysical analyses have provided powerful new approaches to examine the insect–microclimate interface (Porter and Gates, 1969; Anderson et al., 1979).

A useful experimental technique when examining heat exchange in insects in nature is to determine the operative environmental temperature (T_e) experienced by the insects (Bakken and Gates, 1975; Bakken, 1976a). This technique utilizes a model of the animal which matches its external characteristics, including surface area, texture, and coloration. Since the model produces no heat, its equilibrium temperature (T_e) will be determined solely by the environment. This method has been employed to examine heat exchange in ectothermic and endothermic vertebrates, and also recently to examine insect behaviour and thermoregulation (Chappell, 1982, 1983; May, 1982; Joos et al., 1986). Using T_e provides a realistic estimate of the actual environmental conditions faced by a given animal. Differences in temperature between a live, active animal and a dead model are the result of physiological control of heat gain and/or heat loss. This difference is referred to as the physiological offset temperature ($T\Delta$), and for endotherms $T_b = T_e + T\Delta$ (Bakken and Gates, 1975). For strict ectotherms, the measured T_e should equal the animal's T_b unless there is physiological heat transfer. Using models ("T_e thermometers") it is possible to map the entire range of thermal conditions in the habitat. These models are more effective than other measurements of the microclimate because they match the surface characteristics of the animal, thereby integrating the effects of radiation and convection. The T_e approach is attractive because it reduces the calculations and assumptions involved in predicting the thermal environment (see Bakken, 1976a, for detailed discussion).

Mathematical energy balance models usually utilize a "lumped parameter" analysis, where body temperatures, operative temperatures and conductance are averaged over the entire body. However, the different tagmata of adult insects vary extensively in size, surface to volume ratio and insulation (Church, 1960; Heinrich, 1974a). The temperatures of various body regions can vary extensively, particularly during flight (see below). A complete analysis of body temperatures and physiological control of heat exchange from different body regions requires a distributed parameter analysis, where avenues of heat exchange are quantified for each body region (Fig. 1). Even in the absence of

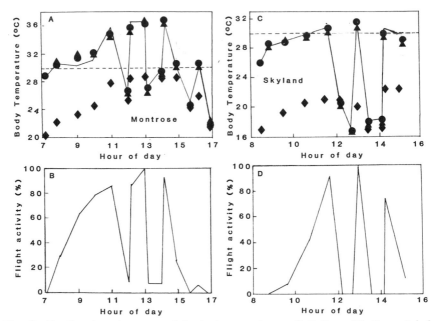

Fig. 2 Predicted and measured body temperatures of *Colias philodice eriphyle* butterflies in two different habitats (A and C). Trianges and circles indicate measured basking temperatures of two individual males throughout the day. Solid line indicates predicted temperature based on heat balance model. Diamonds indicate air temperature. Dashed line indicates minimum flight temperatures. Observed levels of flight activity from two different habitats (B and D). (Modified from Kingsolver, 1983a.)

metabolic heat production, significant differences in temperature can occur in different regions of the insect's body in the field (Heinrich and Casey, 1978). Temperature differences between body regions are crucial for the physiological mechanisms of heat transfer (see below).

From a physiological standpoint, heat budgets are useful for evaluating the relative importance of a particular mechanism of heat transfer such as evaporation (Cooper *et al.*, 1985), or the relative amount of heat exchange occurring in a particular part of the body, in order to determine its importance as a site of regulation (Hegel and Casey, 1982). Heat budgets also allow a quantitative evaluation of the effectiveness of various postures (Casey and Hegel, 1981; Polcyn and Chappell, 1986). For endothermic insects during flight, quantifying the total heat exchange provides an independent method of estimating the metabolic rate of the insect (Weis-Fogh, 1964). Finally, quantifying heat exchange can allow an accurate prediction of the insect's body temperature in the natural environment (Chappell, 1982; Kingsolver, 1983a). Since many activities are strongly temperature-dependent, such predictions are useful in explaining observed activity patterns and evaluating adaptive strategies.

Studies by Kingsolver and colleagues demonstrate the power of heat budget analyses when they are coupled with field behaviour and microclimate measurements. Studying *Colias* butterflies in the laboratory, and using laboratory-derived values for radiant absorbtivity and convection coefficients (Kingsolver and Moffat, 1982), the body temperatures of butterflies in the field could be predicted, based on measured values of radiation and convection (Kingsolver, 1983a). Simultaneous observation of thermocouple-implanted models verified the predicted body temperatures to within 1·5 to 2·0°C. Because flight in these butterflies is limited to thoracic temperatures between 30–40°C (Watt, 1968, 1969), Kingsolver was able to predict the cumulative flight activity time (KFAT) for the butterflies based on microclimate measurements and his predictive climate–space model (Porter and Gates, 1969). Since microclimatic conditions vary as a function of altitude, he demonstrated that fecundity in certain habitats was directly associated with KFAT, which in turn was related to body temperature differences based on solar radiation and wind conditions (Kingsolver, 1983b). Further studies examined "optimal" phenotypes by manipulating the major morphological factors in the climate–space model and compared these findings with measured values for the phenotypic characters along an altitudinal gradient (Kingsolver and Watt, 1983). They found significant differences between the measured values as a function of altitude and close agreement between the predicted optimum and the measured values.

3 Temperature and physiological performance

A variety of studies of insect muscle physiology both *in vivo* and *in vitro* indicate that muscle performance is strongly temperature-dependent (Josephson, 1981). Physiological properties exhibiting the highest thermal sensitivity are those associated with temporal characteristics (contraction frequency, twitch duration, recovery time, etc), while those properties associated with tension development may be relatively or even completely independent of temperature over a wide range. Thus, variation in maximal power output (a combination of these parameters) is usually temperature-dependent. Since the flight muscles are usually the only site of significant metabolic heat production—except in singing insects (Josephson, 1981)—an understanding of muscle function is useful for the evaluation of endothermy in insects. During the pre-flight warm-up of asynchronous fliers (honeybees and bumblebees), the major changes in muscle metabolism are correlated with a strong increase in the frequency of muscle contraction as temperature increases (Heinrich, 1974a). Measurements of action potential frequency correlate directly with changes in metabolic rate (Kammer and Heinrich, 1974). The cost per action potential is comparable over a range of temperatures during warm-up and free hovering flight, indicating that power is changing directly with frequency (Kammer, 1981). A qualitatively similar pattern is shown for several moth species (synchronous fliers) during warm-up and flight (Heinrich, 1971a; Heinrich and Bartholomew, 1971; Casey, 1981b; Casey and Hegel-Little, 1987).

While maximum rates of energy expenditure are highly temperature-dependent, other important physiological functions may be less sensitive. For example, the development of larval insects is temperature-related (Scriber and Slansky, 1981). However, several studies indicate that various aspects of physiological performance, including resting metabolic rate (Casey, 1977; Casey and Knapp, 1987), feeding rates (Sherman and Watt, 1973) and growth rates (Reynolds and Nottingham, 1985; Knapp and Casey, 1986) are thermally independent at body temperatures typical of the prevailing range of air temperatures experienced by the animals. The T_bs maintained by various insect species are those which allow them to maximize a variety of important physiological processes (Heinrich, 1977, 1981b). However, it is not always obvious that an insect needs to regulate its temperature and, if so, how precise that regulation must be.

4   Temperature regulation during flight

Control of thoracic temperature (T_{th}) during flight occurs in a wide variety of insects from several taxa. Due to the thermal dependence of flight muscles, a minimum flight temperature must often be achieved to insure that the insect is capable of prolonging enough lift to sustain its weight. The level of heat production of some flying insects is so great that overheating would follow if T_{th} was not actively regulated physiologically (Heinrich, 1974a). Flight metabolism is independent of environmental temperature in a wide variety of moths (Heinrich, 1971a; Casey, 1976b, 1980, 1981b), bumblebees (Heinrich, 1975), honeybees (Heinrich, 1980b) and solitary bees (Chappell, 1984). The level and constancy of T_{th} over a range of environmental temperatures is related to size; this is due to increases in thoracic conductance and increased levels of heat loss as size decreases (Bartholomew and Epting, 1975; Bartholomew, 1981).

In free flight, the metabolic rate varies widely between different insects and the costs are associated with both the size and shape of the insect. Differences in metabolic rates are associated with differences in wing morphology in sphingid and saturniid moths (synchronous fliers) of a given body mass (Bartholomew and Casey, 1978). Sphingids have smaller wings than saturniids relative to their size and must operate at higher wingstroke frequencies (and therefore at flight muscle contraction frequencies). Differences in metabolic rates reflect different wingstroke frequencies. The mass-specific power input per wingstroke of moths, regardless of their size or wing morphology, is generally similar (Casey, 1981c). Within groups of bees (asynchronous fliers), a similar pattern emerges. The metabolic rates of different Euglossine bees are directly correlated with their respective wingstroke frequencies (Casey et al., 1985). Muscle mass-specific power input per wingstroke is similar, regardless of morphology, for different groups of Euglossine bees (Casey, Ellington and Gabriel, 1987). Synchronous and asynchronous fliers differ in the magnitude of energy expended per wingstroke per unit of flight muscle mass and this difference may be associated with differences in contraction dynamics of the fibrillar muscles (Ellington, 1985). Variation in morphology is associated with variation in stroke frequency which is directly related to metabolic rate and, as a result, to heat production.

The heat produced during flight is a byproduct of the flight effort and represents that portion of the energy expended by the flight muscle which is not converted into useful work. Since muscle is inefficient, most of the energy expenditure during flight is degraded to heat. Physiologists routinely assume a mechanical efficiency of about 20%. However, Weis-Fogh and Alexander (1977) calculated that the mechanical characteristics of flight muscle yield a maximum mechanical power output of about $250\,W/Kg$ muscle, a small fraction of the metabolic rate of animals in hovering flight. Based on a comparison of aerodynamic power requirements and measured metabolic rates of insects in free hovering flight, Ellington (1984, 1985) determined that the mechanical efficiency of synchronous and asynchronous fliers is in the order of about 5–8% of the metabolic rate. Assuming that the flight muscles are the only significant site of energy expenditure during flight, heat production represents in excess of 90% of the total energy expenditure.

The cost of flight, and therefore the level of heat production, is determined by aerodynamic configuration which, in turn, determines a minimum wingbeat frequency. Since muscle frequency is temperature-dependent (Josephson, 1981), either basking or pre-flight warm-up is often required to achieve a minimum wingstroke frequency, unless the wings are so large that only very low frequencies are required (e.g. geometrid moths, Bartholomew and Heinrich, 1973; Casey and Joos, 1983). Once the thoracic temperature for the minimum wingstroke frequency is reached, the animals are capable of flight and the level of heat production determines whether they will overheat, produce a constant temperature excess ($T_{th} - T_a$) over a range of T_as (Weis-Fogh, 1956; Heinrich, 1974a), or have T_{th} indistinguishable from T_a. However, during flight, metabolism and performance appear largely independent of thoracic temperature. Gypsy moths fly over a T_{th} range of at least $22°C$

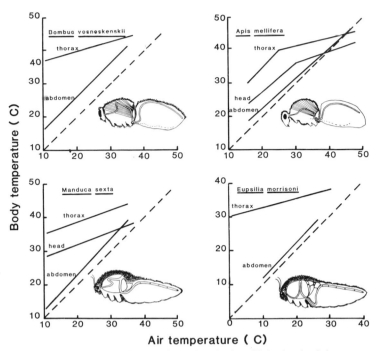

Fig. 3   Body temperatures of several species during flight in the laboratory over a range of air temperatures. (*a*) Bumblebee (from Heinrich, 1976); (*b*) honeybee (Heinrich, 1980a, b); (*c*) the sphinx moth *Manduca sexta* (Hegel and Casey, 1982); (*d*) a winter flying noctuid moth (Heinrich, 1987).

(Casey, 1980), honeybees fly at T_{th}s of 30–45°C (Heinrich, 1980b; Fig. 3), and bumblebees (Heinrich, 1975) and some dragonflies (May, 1976a) show similar ranges of T_{th}. During tethered flight, oxygen consumption and wingstroke frequency in the tropical dipteran, *Pantophthalmus tabininsus* was largely independent of T_{th} (Bartholomew and Lighton, 1986). These data are generally consistent with several other studies which indicated that although the muscle frequency and oxygen consumption of insects during pre-flight warm-up are strongly temperature-dependent (Bartholomew, Vleck and Vleck, 1981), during flight both parameters are relatively temperature-insensitive (May, 1981).

Differences in flight morphology are correlated with differences in the capacity for, and the necessity of thermoregulation during flight, as well as with flight performance and the range of T_{th} over which the insect operates. For example, the tent caterpillar moth, *Malacosoma americanum* has relatively small wings which must operate at a high wingstroke frequency in order to produce the required lift. As a consequence, an obligatory period of pre-flight warm-up is required to raise T_{th} prior to take-off (Casey *et al.*, 1981). During flight this species regulates T_{th} between 35 and 42°C at T_as of 10–30°C. The gypsy moth (*Lymantria dispar*) is similar to the tent caterpillar moth in size and thoracic conductance but has much larger wings. It operates at a much lower wingstroke frequency as a result of its lower wing loading (body mass–wing surface area). In the laboratory, the T_{th} of gypsy moths is about 7°C above T_a, regardless of T_a (Casey, 1980). However, its lower power requirements and low stroke frequencies allow it to operate at considerably lower T_{th} and thermoregulation does not appear to be necessary. The geometrid moth *Operophtera bruceata* routinely flies in the winter at air temperatures near zero. The thoracic temperature of this species is barely above T_a. Since it flaps only a few times per second, it can generate sufficient lift even at a T_{th} near 0°C by virtue of its unusually low wing loading. In fact this species is capable of flight over a 25°C range of T_{th} (Heinrich and Mommsen, 1985).

Although geometrid moths vary dramatically from noctuid moths in their thermoregulation and in the range of muscle temperatures at which they can operate (Bartholomew and Heinrich, 1973; Casey and Joos, 1983; Heinrich and Mommsen, 1985), calculated metabolic rates for the geometrids during flight are similar to the rates exhibited by noctuids during pre-flight warm-up at the same muscle temperature. Moreover, thermal characteristics of citrate synthase and pyruvate kinase enzymes from flight muscle in a geometrid, a noctuid and the sphinx moth, *Manduca sexta*, were very similar, indicating that adaptations at the enzyme level are highly conservative while changes in morphology affecting flight performance and thermoregulation vary radically (Heinrich and Mommsen, 1985).

Several studies have attempted to quantify all the avenues of heat exchange during flight. When calculated rates of heat loss are similar to measured rates of heat production, an independent physiological assessment of flight metabolism has been attained. Until recently this was the only method available for estimating the metabolic rate of insects during forward flight. Weis-Fogh (1964) used this approach to determine the flight energetics of locusts on a flight balance. Since the locusts did not regulate T_{th}, but rather produced a constant temperature excess regardless of T_a, these data probably represent a reasonable estimate of energy expenditure under the conditions of lift and thrust measured. However, verification by comparison with measured heat production is desirable. A similar budget for gypsy moths during free flight slightly underestimates total heat loss compared with measured rates of heat production (Casey, 1980).

For thermoregulating insects, the calculation of heat loss from measurements of T_{th} and thoracic conductance alone will not yield reasonable estimates of the flight metabolism. This is because firstly, at high T_a the driving force for thoracic heat loss ($T_{th} - T_a$) declines while heat production is independent of T_a and secondly, because different parts of the body lose heat at variable rates because their temperature can be controlled by the amount of warm blood which is shunted to that part of the body (Heinrich, 1971a, b, 1976). A heat balance sheet constructed for the sphinx moth *Hyles lineata*, to account for heat loss from the thorax, abdomen and via respiratory evaporation (Casey, 1976b), confirmed Heinrich's conclusion about the importance of the abdomen as a thermal window, but only accounted for 60–70% of the total measured heat production. Since then, the importance of the head as a site of heat transfer has become apparent. In the sphinx moth, *Manduca sexta*, head temperature during free flight can exceed air temperature by as much as 15°C and closely follows thoracic temperature (Hegel and Casey, 1982; Fig. 3). These data suggest that head temperature is regulated. Comparison of calculated heat loss during free flight agrees with measured heat production within 10% at low and intermediate T_a, indicating the large magnitude of heat exchange in the head (Hegel and Casey, 1982).

Heinrich's (1970, 1971a, b) now classic work on *Manduca sexta* documented regulation of heat loss by control of blood circulation (Fig. 3). Subsequently, he reported a similar response in bumblebees (Heinrich, 1976) and elegantly demonstrated the mechanisms of heat transfer. In bumblebees, transfer of blood between thorax and abdomen is prevented by a counter-current exchanger in the petiole. Boluses of warm blood passing from the thorax transfer much of their heat to boluses of cool blood from the abdomen which arrive at the petiole simultaneously thereby sequestering most of the heat in the thorax. During conditions of potential heat stress, this exchanger can be physiologically bypassed. Boluses of blood travelling in opposite directions between thorax and abdomen traverse the waist alternately rather than simultaneously, thereby minimizing heat transfer. Both the ventral diaphragm and the dorsal vessel produce coordinated pulsations (Heinrich, 1976). A similar mechanism appears to operate in carpenter bees (Chappell, 1982; Heinrich and Buchman, 1986). Several large dragonflies also exhibit the capacity to transfer large quantities of heat to the abdomen (May, 1976a); ligation of the dorsal vessel in the abdomen abolishes this transfer (Heinrich and Casey, 1978). A similar mechanism of physiological heat transfer occurs in light-seeking robberflies (Morgan *et al.*, 1985). In contrast, perching dragonflies, which are primarily ectothermic, do not appear to be capable of such heat transfer (Heinrich and Casey, 1978).

For the abdomen to serve as an effective thermal window, the mean abdominal temperature difference ($T_{ab} - T_{e(ab)}$) must increase with increasing T_a in order to offset the reduction in thoracic heat loss. This response has been demonstrated in the laboratory for bumblebees (Heinrich, 1975) and for sphingids (Casey, 1976b; Hegel and Casey, 1982). However, this pattern was not found in a number of flying diurnal insects in the field, including bumblebees (Heinrich, 1972), carpenter bees (Chappell, 1982; Baird, 1986), Euglossine bees (May and Casey, 1983) and dragonflies (M. L. May, personal communication). In all of these cases, when T_{ab} is plotted against T_a, the slopes of the relationship are one or less. Enhanced heat loss from the abdomen should result in a slope of T_{ab} versus T_a of greater than one, indicating that the driving force for abdominal heat loss increases as T_a increases (Heinrich, 1975). These data suggest either that regional thermal gradients within the abdomen are too complicated to accurately assess abdominal heat loss based on a single measurement of T_{ab}, or that additional regulated sites of heat exchange need to be evaluated.

Regulation of T_{th} need not require the use of the abdomen as a thermal window. In honeybees, thoracic temperature varies considerably at lower T_as and becomes progressively more constant as T_a increases (Heinrich, 1980b; Fig. 3). Their small size and relatively poor insulation suggest that, under most circumstances honeybees need to minimize extrathoracic heat loss. The dorsal vessel at the point of contact between the thorax and the abdomen contains a large number of loops; these tend to maximize counter-current exchange between the thorax and the abdomen (Fig. 3). As a consequence of the morphology of the dorsal vessel, the capacity for heat transfer from the thorax to the abdomen is eliminated in honeybees (Heinrich, 1980a, b). The relatively tight control of thoracic temperature at high T_as in honeybees indicates that a different mechanism is operating. As T_a increases, the effectiveness of an abdominal heat exchanger would be sharply reduced because the driving force for heat exchange is approaching zero. The only mechanism for losing large quantities of heat under these circumstances is by evaporation. Honeybees regurgitate fluid from their honeystomachs and evaporate it from their mouthparts. This causes the head temperature to drop due to evaporative cooling and increases the temperature gradient between the thorax and the head. Heat is transferred passively from the thorax by conduction and actively via blood circulation to the head, and from there is dissipated to the environment. This novel mechanism allows honeybees to forage at thoracic temperatures in excess of 45°C (Heinrich, 1980b; Cooper et al., 1985). For most insects, the amount of water required to maintain T_b below T_a by evaporative cooling is prohibitive (Edney, 1977). In honeybees, the water content of bees foraging on nectar is high and the nectar load must be evaporated in the honeymaking process.

The mechanisms of evaporation and convection of heat from the head appear to be operating at cross purposes. Since the driving force for convective heat exchange between the head and the environment is the temperature difference, increasing $T_h - T_a$ would be an effective way of maximizing the convective heat exchange from the head. However, evaporation from the head would reduce the head temperature, thereby reducing the effectiveness of convective heat exchange. Evaporative cooling appears to occur only at very high T_a where a large temperature difference between the head and the environment is not possible. At lower T_a, convection from the head may be important. Further studies of evaporative heat exchange are needed at high temperatures during flight, both to quantify the magnitude of evaporation and to determine its relative importance as a mechanism of regulating heat exchange from a comparative viewpoint.

4.1 HEAD TEMPERATURE

Several recent studies have examined the head temperatures of endothermic insects during flight. Heinrich (1979a, 1980a, b) showed that the head temperatures of honeybees were not regulated at low T_a, but represented the major site of heat loss at very high T_as. Honeybees deliver heat from the thorax via blood circulation and most of this heat is dissipated by the evaporation from the bees mouthparts of a droplet of fluid regurgitated from the honeystomach. Cooper et al.(1985) confirmed these findings for honeybees foraging in the desert and suggested that foraging for nectar may occur preferentially at high T_as when there is a necessity for evaporative cooling. In contrast, in the sphinx moth *Manduca sexta*, head temperature appears to be regulated by means of controlled blood circulation between the head and the thorax (Hegel and Casey, 1982). During flight, head temperature is regulated at least as well as T_{th} (Hegel and Casey, 1982; Fig. 3). If, as in honeybees, evaporative cooling occurs at the mouthparts at high T_as (Adams and Heath, 1964), the effect would be to reduce T_h (thereby more effectively regulating it), while additional heat could be transferred from the thorax. This may explain why heat budgets at high T_a underestimate heat production (Casey, 1976b), but data to evaluate this suggestion are not available.

Baird (1986) reported that regulation of head temperature occurs in the carpenter bee *Xylocopa virginica*, and that differences in patterns of regulation occur between males and females which are consistent with their respective flight patterns. Males hover extensively while waiting for females and regulate T_h precisely, while the females regulate less well. The mechanism of regulation has not been examined. The possible adaptive significance of this difference may be associated with increased visual acuity for the males, thereby increasing the probability of reproductive success (Baird, 1986).

Elevated head temperatures have been reported in several other species. During pre-flight warm-up, the tent caterpillar moth *Malacosoma americanum* elevates T_h by up to as much as 10°C above the surrounding air temperature (under negligible ambient radiant head load), and this increase directly follows the thoracic temperature excess (Casey et al., 1981). While elevated head temperatures in *M. americanum* are associated with active transfer of heat, it is not clear whether this species actually regulates head temperature. In certain noctuid moths, which fly in winter at environmental temperatures near 0°C, the head temperature is much less elevated (Heinrich, 1987). In view of the direction of blood flow (posterior to anterior) in the dorsal vessel, some convective heat transfer of warm blood to the head is inevitable. However, a counter-current heat exchanger developed as a result of close association of the anterior and posterior portions of the vessel as it travels through the thorax may reduce "leakage" of heat from the thorax to the head while allowing normal blood flow (Fig. 3). It is postulated that very low environmental temperatures provide the selective pressures associated with this mechanism of sequestering heat in the thorax (Heinrich, 1987). However, since data are not presented for head temperature during free flight at low temperatures, further studies are needed to evaluate this suggestion. In Euglossine bees, head temperatures may exceed T_a by as much as 10°C. Since these animals routinely fly in shade, it is unlikely that T_e is sufficiently high to account for this difference and physiological augmentation of T_h is suggested. However, it is unclear whether head temperature is regulated in the Euglossines during flight (May and Casey, 1983).

4.2 VARIATION OF FLIGHT PERFORMANCE

An evaluation of body temperatures, heat exchange and flight metabolism in nocturnal insects during flight have been relatively straightforward because the flight muscles represent the only major site of heat gain. For diurnal insects, solar radiation represents a substantial, and often necessary, additional heat source, making quantitative interpretation of flight performance based on heat exchange more difficult. For many insect species, particularly perching dragonflies (May, 1976a; Heinrich and Casey, 1978), dipterans (Heinrich and Pantle, 1975; Gilbert, 1984; Morgan *et al.*, 1985), and butterflies (Watt, 1968, 1969; Rawlins, 1980; Kingsolver, 1983a, b; Heinrich, 1986a, b), solar radiation represents the major source of heat for elevated T_{th} during flight. Consequently, quantitative flight performance estimates in the field become complicated by assumptions associated with radiative heat exchange (Chappell, 1982; Cooper *et al.*, 1985).

In general, the independence of energy metabolism and ambient temperature leads to the conclusion that regulation of body temperatures must be accomplished by regulation of heat loss (Heinrich, 1974a; Kammer, 1981). Dragonflies may regulate their heat production during flight by variations in flight performance (May, 1981). At high T_as the animals spend a larger proportion of their time gliding than at lower T_as where flight was continuously powered (May, 1983). This would reduce metabolic heat production, thus reducing the chances of overheating. Unwin and Corbet (1984) were the first to produce data on the measured wingstroke frequency of bees and flies during free flight over a range of T_as in the field. In the larger bee species, wingstroke frequency tended to increase as temperature decreased, suggesting that the level of heat production (and presumably performance, i.e. flight speed) increased in response to temperature. Unfortunately, the authors had no information on either thoracic temperature or on flight speed, so it is unclear what mechanisms are involved. Nevertheless, their data do suggest variations in flight performance in response to thermal conditions. Small species exhibited increased wingstroke frequencies with increased temperatures which would be expected if they were not regulating their body temperature.

Due to small size and high surface to volume ratio, convective heat exchange in flying insects is high and is relatively sensitive to wind speed (Church, 1960). During basking, convective heat exchange is probably relatively low due to boundary layer effects associated with the substrate on which the insect is perched. In flight, forced convection occurs not only as a result of air speed, but probably as a result of wing movement. In syrphid flies, although heat production during flight is presumably greater than during perching, thoracic temperature declines immediately after take-off (Gilbert, 1984). Similarly, in small butterflies, thoracic temperature during flight is consistently lower than T_{th} during basking in the same environment (Heinrich, 1986b).

Carpenter bees fly throughout the day in the deserts of south-western USA, despite high ambient temperatures and intense solar radiation. They maintain thoracic temperatures of 39–46°C at air temperatures of 20–40°C. Solar radiation has a strong effect on T_{th}, but only at low wind speeds. The T_e of a bee in full sunlight is 8–10°C above air temperature, but that difference drops to 2°C or less in moderate to high winds (Chappell, 1982). These results suggest that hovering flight was only possible for short periods due to intense exogenous and endogenous heating. In order to remain airborne during certain periods, fast forward flight and consequent convective cooling were required. Carpenter bees chose to fly 20–30 metres above the ground, presumably to get away from warmer air near ground level (Chappell, 1982). Another recent study on carpenter bees could not demonstrate conclusively the mechanism for regulation at high T_as and suggested that variations in flight speed and the resultant convective cooling might be necessary (Heinrich and Buchmann, 1986). This conclusion depends on the effect of metabolism at different flight speeds, since additional metabolism (and consequent heat production) could offset increased convective cooling at higher flight speeds (Ellington et al., 1986, and in preparation) measured oxygen consumption of free flying bumblebees over a range of speeds from 0–4 m/s and found that flight metabolism is independent of flight speed. Assuming that this relation holds for other insects, variations in flight speed could be used to modify the rate of heat loss by convection without affecting the rate of heat gain.

5 Endothermy and non-flight activity

During flight, the rate of heat production can not be easily varied for the purposes of thermoregulation because the muscles are operating at the level required to produce lift. When at rest, the large majority of insects are ecto-thermic. However, there are circumstances when an insect is not flying and where it is advantageous to remain endothermic. Unlike most birds and mammals which are primarily endothermic (although there are conspicuous exceptions—see Bartholomew, 1981), or reptiles and amphibians which are entirely ectothermic, some insects maintain elevated thoracic temperatures as a byproduct of muscle contraction. This "intermittant endothermy" serves to enhance performance and, unlike the flight effort where metabolic requirements are essentially fixed, the heat production of the flight muscles can be regulated according to environmental demands.

Male tettigoniids must warm up their muscles prior to singing. *Euconocephalus rhobustus* must operate at muscle temperatures of 30–35°C to produce the appropriate song, and this species regulates muscle temperature over a range of T_a (Josephson, 1973). In the bladder cicada, warming of the tymbal occurs as a consequence of singing and regulation of muscle temperature is not apparent (Josephson and Young, 1979).

Brooding bumblebees incubate their broods by transferring warm haemolymph from the thorax to the abdomen where it is transferred to the brood by conduction (Heinrich, 1974b). During this activity, the sole purpose of the muscle contractions is to produce heat. The rate of heat production in these circumstances is strongly inversely related to ambient temperature. Unlike flight, rates of heat production are varied in relation to ambient temperature in order to control the brood temperature.

Large beetles can maintain high thoracic temperatures in the absence of any other activity. Bartholomew and Casey (1977a) reported maintenance of high, relatively stable T_{th} in tropical scarab beetles; they noted that the elevation of T_{th} occurred cyclically and coincided with abdominal pumping movements, associated with the increased ventilation required to oxygenate the working muscles. They also demonstrated that the aerobic scope of larger beetles was substantially greater than that of smaller beetles and this greater scope coincided with an increased muscle temperature (Bartholomew and Casey, 1977b).

Studies of the African dung beetle (Bartholomew and Heinrich, 1978; Heinrich and Bartholomew, 1979) documented considerable endothermy and indicated that competitive ability was enhanced in beetles exhibiting higher thoracic temperatures. Warmer beetles won competitions for dung 90% of the time, were more effective at making dung balls, and rolled them away faster. Morgan and Bartholomew (1982) reported that scarab beetles maintained a given T_{th} by increasing oxygen consumption in response to decreasing air temperatures.

The level and constancy of regulation of thoracic temperature is often closely associated with the energy supply in the environment. For example, foraging bumblebees vary their thoracic temperature regulation according to the nectar reward provided by the flowers that they are utilizing. When harvesting low nectar flowers such as goldenrod, the bees land on the blooms and walk between blossoms; during this, they allow their thorax temperature to cool to ambient. When utilizing an energy rich food source such as meadowsweet, the bees regulate T_{th} rather precisely (Heinrich, 1972). Under circumstances of reduced competition, even conspicuously endothermic flyers may adopt a "lackadaisical" foraging strategy. Although flower scarabs (*Pachynoda sinuata*), are capable of pre-flight warm-up in the laboratory without radiant heating, in the field they allow their body temperatures to drop near ambient immediately after landing on a flower and usually do not initiate pre-flight warm-up unless it is augmented by solar radiation (Heinrich and McClain, 1986).

6 Ectothermy

A great deal of work has been conducted on large, conspicuously endothermic insects, but far more insects appear to be either routinely or entirely ectothermic. This simplifies the analysis of their heat exchange, reducing the major avenues to radiation and convection (Parry, 1951; Digby, 1955). General patterns of behavioural thermoregulation (i.e. the importance of orientation, postural adjustments, microhabitat selection) are well documented and will not be discussed here. Casey (1981a) and May (1983) provide detailed discussions on these subjects. Butterflies are highly conspicuous and their thermoregulatory behaviour has been examined frequently (Casey, 1981a). Differences in basking postures have resulted in the classification of basking behaviour according to wing position. Dorsal baskers such as Monarch butterflies hold their wings out horizontally (Rawlins, 1980). Lateral baskers such as *Colias* hold their wings vertically closed (Watt, 1968), and rotate the body (yaw and roll angles) to keep the wing surface broadside to incoming radiation. In both cases, the major radiant heating occurs at the

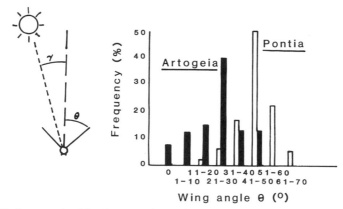

Fig. 4.   Reflectance basking by *Pieris* butterflies (*a*) Basking posture in which butter-flies control the body orientation angle with respect to the sun and the wing angle with respect to the body. (*b*) Frequency distribution of the wing angles during basking in two subgenera *Artogeia* (*n* = 174) and *Pontia*. (After Kingsolver, 1985a.)

wing bases where the degree of melanin pigment is correlated with environ-mental conditions (Watt, 1969; Douglas and Grula, 1978; Kingsolver, 1983a). Satyrine butterflies utilize both dorsal and lateral basking patterns at different times of the day; changing from one pattern to the other may be associated with enhanced visual acuity (Heinrich, 1986a).

K ingsolver (1985a, b) presented a new type of basking in pierid butterflies which he calls reflectance basking (Fig. 4). In this posture, the wings operate as solar reflectors. Sunlight is intercepted and reflected from the wing surface to the dorsal surface of the thorax. Variation in pigment patterns on different parts of the wing (not just the wing base), as well as the angle of the wings relative to the body, increase or decrease the efficacy of this behaviour (Kingsolver, 1985a, b).

Although the role of basking to control the rate of radiant heating is well studied in a variety of ectothermic flying insects (Watt, 1968; May, 1976a; Heinrich and Casey, 1978; Douglas, 1979; Wasserthal, 1975), the role of con-vection cooling in the process has received less attention. Variation in pos-ture and location of the body with respect to wind direction should be as important to the butterfly as radiation because of the importance of convec-tion in the energy balance. In *Colias* butterflies, wind tunnel experiments in-dicate that orientation from 0 to 90° with respect to the prevailing wind has no effect on convective heat exchange, presumably because in the lateral basking posture the wings shield the body (Kingsolver and Moffat, 1982). A recent study by Polcyn and Chappell (1986) examined the body temperatures of butterflies in various basking positions (wings horizontal, at 60° from the body and vertical). These authors corroborated the findings of Kingsolver and Moffat (1982) for the lateral basking posture, but found that there could be important differences in the T_e of their models, depending on wind direction. It is often difficult to maximize radiant heating while at the same time minimizing convective cooling.

The complex interactions of morphology with radiation and convection in the field are being examined. In ectothermic butterflies, insulation and solar absorbtivity are important determinants of equilibrium body temperatures in the field (Kingsolver and Moffatt, 1982). Absorbtivity will have an important effect on the amount of radiant heat gain under any given radiant intensity (Willmer and Unwin, 1981). However, variation in absorbtivity is directly associated with the sensitivity of T_b to wind. Fur thickness decreases sensitivity to wind (Kingsolver and Watt, 1983). These two parameters work together to determine the body temperature of *Colias* spp. butterflies and differences between species and populations occur at different latitudes and altitudes (Kingsolver and Watt, 1983). In addition, pigmentation can change with time, either between instars (Fields and McNeil, 1987) or between generations (Watt, 1969; Douglas and Grula, 1978). In each case, the variation in coloration is consistent with a thermoregulatory function. These studies confirm that the degree of coloration of individuals in populations is closely associated with the microclimate and can be dynamic in both space and time.

For insect larvae, although there are benefits to controlling body temperature, these can be outweighed by heavy predator or parasitoid pressure (Heinrich, 1979b). The setae of many caterpillars are not as thick and dense as on the thorax of moths and bumblebees, but are rather more sparsely distributed. In gypsy moth caterpillars setae are longer and more numerous on the lateral portions of the body. Solar radiation can penetrate and directly warm the body surface, while the effective diameter of the caterpillar is increased, thereby retarding convective heat exchange (Casey and Hegel, 1981). In Arctic woolly-bear caterpillars, although the setae are more uniformly distributed, a similar mechanisms occur (Kevan *et al.*, 1982; Fields and McNeil, 1987).

Very small size reduces the ability to achieve elevated body temperatures (Bartholomew, 1981; Stevenson, 1985a, b). While a variety of caterpillars control their temperatures behaviourally, the temperature excesses achieved and the precision of regulation is modest (Sherman and Watt, 1973; Casey, 1976a; Rawlins and Lederhouse, 1981). Eastern tent caterpillars combine tent architecture, which allows penetration of solar radiation while reducing convective cooling, together with behavioural aggregation to further reduce individual surface to volume ratio. In early April, tent caterpillars weighing only 10–20 mg exhibit aggregation temperatures as high as 20°C above T_a (Knapp and Casey, 1986). This is more than ten times the temperature excess

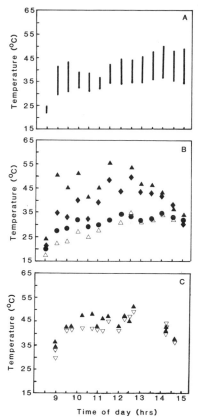

Fig. 5   Thermal relations of eastern tent caterpillars in the field. (*a*) The range of air temperatures in different locations inside the tent. (*b*) Operative temperatures (T_e) of models placed in various locations (closed circles = underside of tent (in shade); diamonds = single model on dorsal tent surface in sunlight; closed triangles = "clump model" on the same dorsal surface consisting of five freeze dried caterpillars attached to resemble a natural caterpillar aggregate; open triangles = a single caterpillar model placed on a branch away from the tent. (*c*) Body temperatures of living caterpillars under the same environmental conditions, utilizing the thermal heterogeneity of the tent microclimate. (Data from Joos *et al.*, 1986.)

predicted for an individual caterpillar of the same size (Stevenson, 1985a). As the size and thermal capacity of the caterpillars and the tent change during the season, the environmental conditions also change. Later in the season, the tent serves as a radiation shield to prevent overheating of the caterpillars. A T_e model of a late instar tent caterpillar on a branch in full sunshine is only a few °C above T_a, while a similar model placed on the surface of a tent can be 30°C above T_a under normal radiant loads (Fig. 5). Clumping together further increases body temperature. Small aggregates of T_e models on the same tent surface may be 6–7°C above the temperature of the individual on the same surface (Joos et al., 1986). The large elevations in body temperature of the models are due to several factors. The tent surface provides a large boundary layer which should reduce convective heat loss. Aggregation further decreases the surface to volume ratio and convection, while still allowing radiant heat uptake. The tent warms several degrees above air temperature, further enhancing the T_b of the caterpillars. In addition to allowing the caterpillars to heat up far above the levels they could achieve without the tent, the tent also provides a range of temperatures which the caterpillars can utilize by microhabitat selection. During the middle of the day the body temperature of caterpillars can differ by as much as 20°C, depending on their location in the tent (Fig. 5). This highly heterogeneous environment allows the caterpillars to regulate their T_b quite precisely (Joos et al., 1986).

7 Concluding remarks

The study of insect thermoregulation has been advancing on a number of fronts. Application of comprehensive heat exchange theory is allowing a more detailed evaluation of the range of possible thermal environments an insect can utilize, as well as providing predictions of body temperatures under a variety of environmental conditions. At the same time, physiological studies are addressing not only the mechanisms of thermoregulation, but also the thermal responses of organs and systems and the range of preferred temperatures required for growth and physiological performance. Careful behavioural studies of particular species, in conjunction with biophysical and physiological studies, are providing a new understanding of the nature and types of adaptations of different insect species as well as a more rigorous and testable series of explanations about the nature of energy exchanges between insects and their environments. While general patterns are clear, the staggering diversity of insects ensures that there will be many different, specific adaptive responses to similar selective pressures.

The thermal biology of insects is fundamental to all their physiological functions. An analysis of the thermal physiology and ecology of the insect, coupled with an understanding of the environment and the insect's behaviour, will reap dividends in excess of explanations of specific phenomena. Integrated, multidisciplinary approaches are providing exciting new opportunities for the synthesis of ideas. As a consequence, future studies will be able to address larger issues in population biology, genetics and evolutionary biology. The intimate relationship between the insect and its environment, and its superb responsiveness to it, provides a unique opportunity for understanding thermal adaptation. Using the tools available, such analyses can produce an overview which is sensitive to the realities of the organism–environment interface which has shaped adaptation and, ultimately, to the direction of evolutionary change.

Acknowledgements

I am pleased to thank Drs G. A. Bartholomew, B. Heinrich, B. Joos, J. G. Kingsolver, M. L. May, K. R. Morgan and D. Polcyn for access to previously unpublished information. I thank Dr B. Joos for useful discussion and critical evaluation of an earlier draft of this manuscript. I also thank Dr Peter Evans and the editorial staff for their infinite patience. Financial support to the author from the National Science Foundation (Grants no. DCB 84-16975 and BSR 85-06740) and the New Jersey Agricultural Experiment Station (Projects no. 08511 and 08337) are gratefully acknowledged.

References

Adams, P. A. and Heath, J. E. (1964). An evaporative cooling mechanism in *Pholus achemon*. *J. Res. Lepid.* **3**, 69–72.

Anderson, R. V., Tracy, C. R. and Abramsky, Z. (1979). Habitat selection in two species of grasshoppers: the role of thermal and hydric stresses. *Oecologia* **38**, 359–374.

Bakken, G. S. (1976a). A heat transfer analysis of animals: unifying concepts and the application of metabolism chamber data to field ecology. *J. theoret. Biol.* **60**, 337–384.

Bakken, G. S. (1976b). An improved method for determining thermal conductance and equilibrium body temperature with cooling curve experiments. *J. therm. Biol.* **1**, 169–175.

Bakken, G. S. and Gates, D. M. (1975). Heat-transfer analysis of animals: some implications for field ecology, physiology and evolution. *In* "Perspectives of Biophysical Ecology" (Eds D. M. Gates and R. B. Schmerl), pp. 225–290. Springer Verlag, Berlin.

Baird, J. M. (1986). A field study of thermoregulation in the carpenter bee *Xylocopa virginica virginica* (Hymenoptera: Anthophoridae). *Physiol. Zool.* **59**, 157–168.

Bartholomew, G. A. (1981). A matter of size: an examination of endothermy in insects and terrestrial vertebrates. *In* "Insect Thermoregulation" (Ed. B. Heinrich), pp. 45–78. John Wiley, New York.

Bartholomew, G. A. and Casey, T. M. (1977a). Endothermy during terrestrial activity in large beetles. *Science* **195**, 882–883.

Bartholomew, G. A. and Casey, T. M. (1977b). Body temperature and oxygen consumption during rest and activity in relation to body size in tropical beetles. *J. therm. Biol.* **2**, 173–176.

Bartholomew, G. A. and Casey, T. M. (1978). Oxygen consumption of moths during rest, pre-flight warm-up and flight in relation to body size and wing morphology. *J. exp. Biol.* **76**, 11–25.

Bartholomew, G. A. and Epting, R. J. (1975). Allometry of post-flight cooling rates in moths: a comparison with vertebrate homeotherms. *J. exp. Biol.* **63**, 603–613.

Bartholomew, G. A. and Heinrich, B. (1973). A field study of flight temperatures in moths in relation to body weight and wing loading. *J. exp. Biol.* **58**, 123–135.

Bartholomew, G. A. and Heinrich, B. (1978). Endothermy in African dung beetles during flight, ball making and ball rolling. *J. exp. Biol.* **73**, 65–83.

Bartholomew, G. A. and Lighton, J. R. B. (1986). Endothermy and energy metabolism of a giant tropical fly. *Pantophthalmus tabaninus* Thunberg. *J. comp. Physiol.* **156**, 461–467.

Bartholomew, G. A., Vleck, D. and Vleck, C. M. (1981). Instantaneous measurements of oxygen consumption during pre-flight warm-up and post-flight cooling in sphingid and saturniid moths. *J. exp. Biol.* **90**, 17–32.

Casey, T. M. (1976b). Flight energetics of sphinx moths: heat production and heat loss in *Hyles lineata* during free flight. *J. exp. Biol.* **64**, 545–560.

Casey, T. M. (1977). Physiological responses to temperature of caterpillars of a desert population of *Manduca sexta* (Lepidoptera: Sphingidae). *Comp. Biochem. Physiol.* **57A**, 53–58.

Casey, T. M. (1980). Flight energetics and heat exchange of gypsy moths during hovering flight. *J. exp. Biol.* **88**, 133–145.

Casey, T. M. (1981a). Behavioural mechanisms of thermoregulation. *In* "Insect Thermoregulation" (Ed. B. Heinrich), pp. 79–113. John Wiley, New York.

Casey, T. M. (1981b). Energetics and thermoregulation of *Malacosoma americanum* (Lepidoptera: Lasiocampidae) during hovering flight. *Physiol. Zool.* **54**, 362–371.

Casey, T. M. (1981c). Insect flight energetics. *In* "Locomotion and Energetics of Arthropods" (Eds C. F. Herreid and C. R. Fourtner), pp. 419–452. Plenum Press, New York.

Casey, T. M. and Hegel, J. R. (1981). Caterpillar setae: insulation for an ectotherm. *Science* **214**, 1131–1133.

Casey, T. M., Hegel, J. R. and Buser, C. S. (1981). Physiology and energetics of pre-flight warm-up in the Eastern tent caterpillar moth *Malacosoma americanum*. *J. exp. Biol.* **94**, 119–135.

Casey, T. M. and Hegel-Little, J. R. (1987). Instantaneous oxygen consumption and metabolic stroke work in *Malacosoma americanum* during pre-flight warm-up. *J. exp. Biol.* **127**, 389–400.

Casey, T. M. and Joos, B. (1983). Morphometrics, conductance, thoracic temperature and flight energetics of noctuid and geometrid moths. *Physiol. Zool.* **56**, 160–173.

Casey, T. M. and Knapp, R. (1987). Caterpillar thermal adaptation: behavioral differences reflect metabolic thermal sensitivities. *Comp. Biochem. Physiol.* (in press).

Casey, T. M., May, M. L. and Morgan, K. R. (1985). Flight energetics of Euglossine bees in relation to morphology and wing stroke frequency. *J. exp. Biol.* **116**, 271–289.

Casey, T. M., Ellington, C. P. and Gabriel, J. M. (1987). Mechanical and energetic limitations to hovering in Euglossine bees. *J. appl. Physiol.* (in press).

Chappell, M. A. (1982). Temperature regulation of carpenter bees (*Xylocopa californica*) foraging in the Colorado Desert of Southern California. *Physiol. Zool.* **55**, 267–280.

Chappell, M. A. (1983). Metabolism and thermoregulation in desert and montane grasshoppers. *Oecologia* **56**, 126–131.

Chappell, M. A. (1984). Temperature regulation and energetics of the solitary bee *Centris pallida* during foraging and intermale mate competition. *Physiol. Zool.* **57**, 215–225.

Church, N. S. (1960). Heat loss and the body temperature of flying insects. Parts I and II. *J. exp. Biol.* **37**, 171–213.

Cooper, P. D., Schaffer, W. M. and Buchmann, W. D. (1985). Temperature regulation of honeybees (*Apis mellifera*) foraging in the Sonoran desert. *J. exp. Biol.* **114**, 1–15.

Digby, P. S. B. (1955). Factors affecting the temperature excess of insects in sunshine. *J. exp. Biol.* **32**, 279–298.

Douglas, M. M. (1979). Hot butterflies. *J. Natural History* **88**, 56–65.

Douglas, M. M. and Grula, J. W. (1978). Thermoregulatory adaptations allowing ecological range expansion by the pierid butterfly, *Nathalis iole* Boisduval. *Evolution* **32**, 776–783.

Edney, E. B. (1977). "Water Balance of Land Arthropods". Springer Verlag, New York.

Ellington, C. P. (1984). The aerodynamics of hovering insect flight. VI. Lift and power requirements. *Phil. Trans. Roy. Soc. Lond.* B **305**, 145–181.

Ellington, C. P. (1985). Power and efficiency of insect flight muscle. *J. exp. Biol.* **115**, 293–304.

Ellington, C. P., Machin, K. E. and Casey, T. M. (1986). Oxygen consumption of bumblebees in free forward flight. *Am. Zool.* **26**, 89.

Fields, P. G. and McNeil, J. N. (1987). The importance of seasonal variation in hair colouration for thermoregulation of *Ctenucha virginica* larvae (Lepidoptera: Arctiidae). *Physiol. Entomol.* (in press).

Gilbert, F. S. (1984). Thermoregulation and the structure of swarms in *Syrphus ribesii* (Syrphidae). *Oikos* **42**, 249–255.

Heath, J. E. and Adams, P. E. (1965). Temperature regulation in the sphinx moth during flight. *Nature, Lond.* **205**, 309–310.

Hegel, J. R. and Casey, T. M. (1982). Thermoregulation and control of head temperature in the sphinx moth, *Manduca sexta*. *J. exp. Biol.* **101**, 1–15.

Heinrich, B. (1970). Thoracic temperature stabilization by blood circulation in a free flying moth. *Science* **168**, 580–582.

Heinrich, B. (1971a). Temperature regulation of the sphinx moth, *Manduca sexta*. I. Flight energetics and body temperature during free and tethered flight. *J. exp. Biol.* **54**, 141–152.

Heinrich, B. (1971b). Temperature regulation of the sphinx moth, *Manduca sexta*. II. Regulation of heat loss by control of blood circulation. *J. exp. Biol.* **54**, 153–166.

Heinrich, B. (1972). Energetics of temperature regulation and foraging in a bumblebee, *Bombus terricola*. *J. Comp. Physiol.* **77**, 49–64.

Heinrich, B. (1974a). Thermoregulation in endothermic insects. *Science* **185**, 747–756.

Heinrich, B. (1974b). Thermoregulation in bumblebees. I. Brood incubation in *Bombus vosnesenskii* queens. *J. Comp. Physiol.* **88**, 129–140.

Heinrich, B. (1975). Thermoregulation in bumblebees. II. Energetics of warm-up and free flight. *J. Comp. Physiol.* **96**, 155–166.

Heinrich, B. (1976). Heat exchange in relation to blood flow between thorax and abdomen in bumblebees. *J. exp. Biol.* **64**, 561–585.

Heinrich, B. (1977). Why have some animals evolved to regulate a high body temperature? *Am. Naturalist* **111**, 623–640.

Heinrich, B. (1979a). Keeping a cool head: honeybee thermoregulation. *Science* **205**, 1269–1271.

Heinrich, B. (1979b). Foraging strategies of caterpillars: leaf damage and possible predator avoidance strategies. *Oecologia* **42**, 325–337.

Heinrich, B. (1980a). Mechanisms of body-temperature regulation in honeybees, *Apis mellifera* I. Regulation of head temperature. *J. exp. Biol.* **85**, 61–72.

Heinrich, B. (1980b). Mechanisms of body-temperature regulation in honeybees, *Apis mellifera*. II. Regulation of thoracic temperature at high air temperatures. *J. exp. Biol.* **85**, 73–87.

Heinrich, B. (1981a). A brief historical survey. *In* "Insect Thermoregulation" (Ed. B. Heinrich), pp. 7–17. John Wiley, New York.

Heinrich, B. (1981b). Ecological and evolutionary perspectives. *In* "Insect Thermoregulation" (Ed. B. Heinrich), pp. 235–302. John Wiley, New York.

Heinrich, B. (1986a). Thermoregulation and flight activity of the Satyr, *Coenonympha inornara* (Lepidoptera: Satyridae). *Ecology* **67**, 593–597.

Heinrich, B. (1986b). Comparative thermoregulation of four montane butterflies of different mass. *Physiol. Zool.* **59**, 616–626.

Heinrich, B. (1987). Thermoregulation by winter-flying endothermic moths. *J. exp. Biol.* **127**, 313–332.

Heinrich, B. and Bartholomew, G. A. (1971). An analysis of pre-flight warm-up in the sphinx moth, *Manduca sexta*. *J. exp. Biol.* **55**, 223–239.

Heinrich, B. and Bartholomew, G. A. (1979). Roles of endothermy and size in inter- and intraspecific competition for elephant dung in an African dung beetle, *Scarabaeus laevistriatus*. *Physiol. Zool.* **52**, 484–496.

Heinrich, B. and Buchmann, S. L. (1986). Thermoregulatory physiology of the carpenter bee *Xylocopa varipuncta*. *J. comp. Physiol.* **156**, 557–562.

Heinrich, B. and Casey, T. M. (1978). Heat transfer in dragonflies: "fliers" and "perchers". *J. exp. Biol.* **74**, 17–36.

Heinrich, B. and McClain, E. (1986). Laziness and hypothermia as a foraging strategy in flower scarabs (Coleoptera: Scarabaeidae). *Physiol. Zool.* **59**, 273–282.

Heinrich, B. and Mommsen, T. P. (1985). Flight of winter moths near 0°C. *Science* **228**, 177–179.

Heinrich, B. and Pantle, C. (1975). Thermoregulation in small flies (*Syrphus* sp.): basking and shivering. *J. exp. Biol.* **62**, 599–610.

Joos, B., Casey, T. M. and Fitzgerald, T. D. (1986). Roles of the tent in behavioural thermoregulation of Eastern tent caterpillars. *Am. Zool.* **26**, 111.

Josephson, R. K. (1973). Contraction kinetics of the fast muscle used in singing by a katydid. *J. exp. Biol.* **59**, 781–801.

Josephson, R. K. and Young, D. (1979). Body temperature and singing in the bladder cicada *Cystoma saundersii*. *J. exp. Biol.* **80**, 69–81.

Josephson, R. K. (1981). Temperature and mechanical performance of insect muscle. *In* "Insect Thermoregulation" (Ed. B. Heinrich), pp. 19–44. New York, John Wiley.

Kammer, A. E. (1981). Physiological mechanisms of thermoregulation. *In* "Insect Thermoregulation" (Ed. B. Heinrich), pp. 115–158. John Wiley, New York.

Kammer, A. E. and Heinrich, B. (1974). Metabolic rates related to muscle activity in bees. *J. exp. Biol.* **61**, 219–227.

Kammer, A. E. and Heinrich, B. (1978). Insect flight metabolism. *Adv. Insect. Physiol.* **13**, 133–228.

Kevan, P. G., Jensen, T. S. and Shorthouse, J. D. (1982). Body temperatures and behavioural thermoregulation of high arctic woolly-bear caterpillars and pupae (*Gynaephora rosii*, Lymantriidae: Lepidoptera) and the importance of sunshine. *Arctic Alpine Res.* **14**, 125–136.

Kingsolver, J. G. (1983a). Thermoregulation and flight in *Colias* butterflies: elevational patterns and mechanistic limitations. *Ecology* **64**, 534–545.

Kingsolver, J. G. (1983b). Ecological significance of flight activity in *Colias* butterflies: implications for reproductive strategy and population structure. *Ecology* **64**, 546–551.

Kingsolver, J. G. (1985a). Thermal ecology of *Pieris* butterflies (Lepidoptera: Pieridae): a new mechanism of behavioural thermoregulation. *Oecologia* **66**, 540–545.

Kingsolver, J. G. (1985b). Thermoregulatory significance of wing melanization in *Pieris* butterflies (Lepidoptera: Pieridae): physics, posture, and pattern. *Oecologia* **66**, 546–553.

Kingsolver, J. G. and Moffat, R. J. (1982). Thermoregulation and the determinants of heat transfer in *Colias* butterflies. *Oecologia* **53**, 27–33.

Kingsolver, J. G. and Watt, W. B. (1983). Thermoregulatory strategies in *Colias* butterflies: thermal stress and the limits to adaptation in thermally varying environments. *Am. Naturalist* **121**, 32–55.

Knapp, R. and Casey, T. M. (1986). Thermal ecology, behavior, and growth of gypsy moth and Eastern tent caterpillars. *Ecology* **67**, 598–608.

May, M. L. (1976a). Thermoregulation and adaptation to temperature in dragonflies (Odonata: Anisoptera). *Ecol. Monogr.* **46**, 1–32.

May, M. L. (1976b). Warming rates as a function of body size in periodic endotherms. *J. comp. Physiol.* **111**, 55–70.

May, M. L. (1978). Thermal adaptations of dragonflies. *Odonatologica* **7**, 27–47.

May, M. L. (1979). Insect thermoregulation. *Ann. Rev. Entomol.* **24**, 313–349.

May, M. L. (1981). Wingstroke frequency of dragonflies (Odonata: Anisoptera) in relation to temperature and body size. *J. comp. Physiol.* **144**, 229–240.

May, M. L. (1982). Body temperature and thermoregulation of the Colorado potato beetle, *Leptinotarsa decemlineata*. *Ent. Exp. Appl.* **31**, 413–420.

May, M. L. (1983). Thermoregulation. *In* "Comprehensive insect physiology, biochemistry and pharmacology" (Eds G. A. Kerkut and L. I. Gilbert), pp. 507–552. Pergamon Press, Oxford.

May, M. L. and Casey, T. M. (1983). Thermoregulation and heat exchange in Euglossine bees. *Physiol. Zool.* **56**, 541–551.

Morgan, K. R. and Bartholomew, G. A. (1982). Homeothermic response to reduced ambient temperature in a scarab beetle. *Science* **216**, 1409–1410.

Morgan, K. R., Shelly, T. E. and Kimsey, L. S. (1985). Body temperatures regulation, energy metabolism, and foraging in light seeking and shade seeking robberflies. *J. Comp. Physiol. B* **155**, 561–570.

Parry, D. A. (1951). Factors determining the temperature of terrestrial arthropods in sunshine. *J. exp. Biol.* **28**, 445–462.

Pivnick, K. E. and McNeil, J. N. (1985). Sexual differences in the thermoregulation of *Thymelicus lineola* adults (Lepidoptera: Hesperiidae). *Ecology* **67**, 1024–1035.

Polcyn, D. M. and Chappell, M. A. (1986). Analysis of heat transfer in Vanessa butterflies: effects of wing position and orientation to wind and light. *Physiol. Zool.* **59**, 706–716.

Porter, W. P. and Gates, D. M. (1969). Thermodynamic equilibria of animals. *Ecol. Monogr.* **39**, 227–244.

Rawlins, J. E. (1980). Thermoregulation by the black swallowtail butterfly, *Papilio polyxenes* (Lepidoptera: Papilionidae). *Ecology* **61**, 345–357.

Rawlins, J. E. and Lederhouse, R. C. (1981). Developmental influences of thermal behavior on monarch caterpillars (*Danaus plexippus*): an adaptation for migration (Lepidoptera: Nymphalidae). *J. Kans. Ent. Soc.* **54**, 387–408.

Reynolds, S. E. and Nottingham, S. F. (1985). Effects of temperature on growth and efficiency of food utilization in fifth-instar caterpillars of the tobacco hornworm. *J. Insect Physiol.* **31**, 129–134.

Scriber, J. M. and Slansky, F. (1981). The nutritional ecology of immature insects. *Ann. Rev. Entomol.* **26**, 183–211.

Seymour, R. S. (1974). Convective and evaporative cooling in sawfly larvae. *J. Insect Physiol.* **20**, 2447–2457.

Sherman, P. W. and Watt, W. B. (1973). The thermal ecology of some *Colias* larvae. *J. comp. Physiol.* **83**, 25–40.

Stevenson, R. D. (1985a). Body size and limits to the daily range of body temperature in terrestrial ectotherms. *Am. Naturalist* **125**, 102–117.

Stevenson, R. D. (1985b). The relative importance of behavioral and physiological adjustments controlling body temperature in terrestrial ectotherms. *Am. Naturalist* **126**, 362–386.

Tracy, C. R. (1972). Newton's law: its application for expressing heat losses from homeotherms. *Bioscience* **22**, 656–659.

Unwin, D. M. and Corbet, S. A. (1984). Wingbeat frequency, temperature, and body size in bees and flies. *Physiol. Entomol.* **9**, 115–121.

Wasserthal, L. T. (1975). The role of butterfly wings in regulation of body temperature. *J. Insect Physiol.* **21**, 1921–1930.

Watt, W. B. (1968). Adaptive significance of pigment polymorphism in *Colias* butterflies. I. Variation of melanin pigment in relation to thermoregulation. *Evolution* **22**, 437–458.

Watt, W. B. (1969). Adaptive significance of pigment polymorphism in *Colias* butterflies. II. Thermoregulation of photoperiodically controlled melanin variation in *Colias eurytheme*. *Proc. Nat. Acad. Sci. USA* **63**, 767–774.

Weis-Fogh, T. (1956). Biology and physics of locust flight. II. Flight performance of the desert locust. *Phil. Trans. Roy. Soc. Lond.* **B239**, 459–510.

Weis-Fogh, T. (1964). Biology and physics of locust flight. VIII. Lift and metabolic rate of flying locusts. *J. exp. Biol.* **41**, 257–271.

Weis-Fogh, T. and Alexander, R. McN. (1977). The sustained power output obtainable from striated muscle. *In* "Scale effects in animal locomotion" (Ed. T. J. Pedley), pp. 511–526. Academic Press, New York.

Willmer, P. G. (1982). Microclimate and the environmental physiology of insects. *Adv. Insect Physiol.* **16**, 1–57.

Willmer, P. G. and Unwin, D. M. (1981). Field analysis of insect heat budgets: reflectance, size and heating rates. *Oecologia* **50**, 250–255.

Molecular Targets of Pyrethroid Insecticides

David B. Sattelle* and Daisuke Yamamoto†

*A.R.F.C., Unit of Insect Neurophysiology and Pharmacology, Department of Zoology, University of Cambridge, Downing Street, Cambridge CB2 3EJ, U.K. and †Department of Neuroscience, Mitsubishi-Kasei Institute of Life Sciences, 11, Minamiooya, Machida, Tokyo 194, Japan

1   Introduction

The synthetic pyrethroids represent the major innovation in insecticide chemistry in recent years and this class of compounds with diverse agricultural, veterinary and public health applications is dramatically changing the ways in which insect control agents are deployed. A number of excellent reviews of aspects of the chemistry (Elliott, 1971, 1977, 1985; Elliott and Janes, 1973, 1978a, b; Crombie, 1980; Naumann, 1981; Tessier, 1982, 1985) and insecticidal actions (Narahashi, 1971, 1982; Casida *et al.*, 1983; Miller and

ADVANCES IN INSECT PHYSIOLOGY VOL. 20
ISBN 0 12 024220 6

Salgado, 1985) of pyrethroids have appeared in recent years and we shall endeavour not to retread the same path. Rather, we have focussed on the work of the last decade, which deals with the actions of these compounds and sheds light on the molecular targets of pyrethroids.

There are four aspects of insect biology that have proved directly susceptible to the actions of insecticides. These include respiration, cuticle formation and hormone action, but for the great majority of insecticidally-active molecules synthesized to date, the target organ is the insect nervous system (Corbett et al., 1984). Whereas the enzymes for neurotransmitter synthesis and breakdown have hitherto provided the major molecular targets, neurotransmitter receptors, ion channels and membrane transport processes are emerging as candidate targets for the present generation of insecticides, including the pyrethroids.

It is a major goal of many of those working on pyrethroid actions to establish beyond doubt the primary target site of these successful insect control agents. One aim of the present survey is to examine the case for the sodium channel as the primary molecular target. However, the recent literature contains many examples of direct actions of pyrethroids on quite separate sites, both in insects and in other organisms. These other candidate molecular targets will be considered, together with their possible role in the insecticidal actions of pyrethroids. However, insect pharmacology is still in its infancy, so our conclusions in relation to the target organisms will be, at best, a state-of-the-art assessment. Wherever it seems appropriate, we refer to data on pyrethroid actions obtained on other invertebrate and vertebrate preparations. In doing so we have tried to avoid the trap of assuming that the pharmacology of a particular vertebrate receptor or channel will necessarily extrapolate to an insect species containing similar, though perhaps less well-characterized sites.

The importance of pursuing basic studies on pyrethroid molecular targets has been clearly emphasized by Michael Elliott (1985), whose outstanding contributions to pyrethroid chemistry led the way to the major agricultural role for these molecules evident today. In summarizing many years of work by the Rothamsted group, he concluded in retrospect that "no fully rational approach to developing insecticides related to the natural pyrethrins has been possible, primarily because the precise sites and locations of action are still unknown. Surely in the immediate future we should look to more progress there to provide the basis for development of new biologically-active compounds. Combined with detection of activity in lead compounds, especially to be found in natural products, progress should then be much more direct and rational". This represents a major challenge to those working on the biochemistry and physiology of insects. Though at first sight it appears a daunting task, there are some grounds for optimism. Insects are proving suitable experimental material for several aspects of the study of neurotransmitter receptors, ion channels and membrane-bound enzymes. The use of identifiable neurones and pathways in the insect central nervous system is facilitating interpretation of the actions of pharmacological and toxicological agents (including insecticides) monitored by electrophysiological methods. Also, although individual insects provide only small amounts of material for biochemical studies, they can be reared in large numbers to offset this apparent disadvantage. Finally, insect genetics offers experimental approaches to elucidate the structure and functions of potential insecticide molecular targets and the origins of resistance.

Undoubtedly the pyrethroids represent a major resource in the protection of worldwide food supplies. It is to be hoped that their use will be managed with the aim of limiting the development of resistance. Detailed knowledge of the molecular targets of pyrethroids may enable not only the further development of these compounds, but also the initiation of design strategies to overcome resistance problems as they arise.

In this review, a survey of the candidate molecular targets of pyrethroids is preceded by a brief introduction to the chemistry and insecticidal actions of these compounds. Only sufficient detail is included to familiarize the insect biochemist and physiologist with some of their key properties; those seeking further details on the chemistry and applications of pyrethroids are encouraged to consult two excellent multi-authored volumes (Janes, 1985; Leahey, 1985).

1.1   PYRETHROID CHEMISTRY AND SELECTIVE TOXICITY

1.1.1   *Natural and synthetic pyrethroids*

Pyrethroids, the most successful of the present generation of commercial insecticides, have been synthesized based on the chemical structure of natural pyrethrins. This achievement is the outcome of intensive efforts over many years to enhance the biological activity and chemical stability of natural pyrethrins, a mixture of insecticidal esters of plant origin. The relatively recent discovery of photostable synthetic pyrethroids (Elliott, 1977) has led to profound changes in agricultural insecticide technology. These highly active, biodegradable compounds can be used in the field at rates of 30 g/hectare (10–100 fold lower than previously used insecticides). Although relatively costly, their effectiveness, combined with the absence of persistent residues following decomposition in the environment, and their relatively low mammalian toxicity, have been responsible for the widespread current use of pyrethroids. The development of synthetic pyrethroids has involved systematic manipulation of different parts of the molecule and examination of the effects of such changes on biological activity. Structure–activity relations in pyrethroids are based mainly on shape and stereochemistry, as opposed to electronic properties, so that a precise definition in terms of particular chemical groups is not readily achieved. However, if a compound's structure can be reasonably derived from that of the natural pyrethrins and if it shows a range of biological properties that overlap to a considerable degree with those of existing members of the group, then it is commonly referred to as a pyrethroid.

Pyrethrum is prepared by first drying, then crushing to a powder the flowers of *Chrysanthemum cinerariaefolium*; an extract referred to as pyrethrin is prepared by solvent extraction. Discovery of the insecticidal properties of pyrethrum is likely to have taken place in the Caucasus–Iran region of Asia, where the plant is native. Its beneficial effects were known to the Chinese in the first century AD (Lhoste, 1964). Although natural pyrethrins still retain a useful role, they have been surpassed by synthetic analogues, with improved effectiveness and greater diversity of applications than the naturally occurring molecules.

Fujitani (1909) postulated that pyrethrins were esters, but it required another 60 years of research before their absolute stereochemistry was determined. For comprehensive reviews, articles by Elliott and Janes (1973) and Crombie (1980) should be consulted. The six active constituents of pyrethrum extract are esters of two carboxylic acids (chrysanthemic acid and pyrethric acid) and three cyclopentenolones (pyrethrolone, cinerolone and jasmolone). A nomenclature is adopted in which the alcohol component is denoted by name and the acid by number (Fig. 1). Synthesis of analogues of natural pyrethrins stemmed from the demonstration by Staudinger and Ruzicka (1924) that the active components of pyrethrum derived from two carboxylic acids, but details of the first commercially successful synthetic product (bioallethrin) were not published for another 25 years (Schechter *et al.*, 1949). These early synthetic pyrethroids lacked chemical stability and were metabolized by insects. Their primary use was as household insecticides, most often in the presence of a synergist functioning to prevent pyrethroid breakdown (Hodgson and Tate, 1976).

The discovery of the light-stable analogues of pyrethrin has transformed the history of plant protection. With compounds such as cypermethrin, fenvalerate and deltamethrin (= decamethrin), synthetic pyrethroids are rapidly becoming the predominant class of insecticides worldwide. Differences in mammalian and insect toxicity appear to reside in the degree to which these groups of organisms metabolize pyrethroids (Casida, 1983). The enzyme (pyrethroid carboxyesterase) which hydrolyses the esters of chrysanthemic acid was purified from rat liver microsomes. It hydrolyses *trans* isomers of

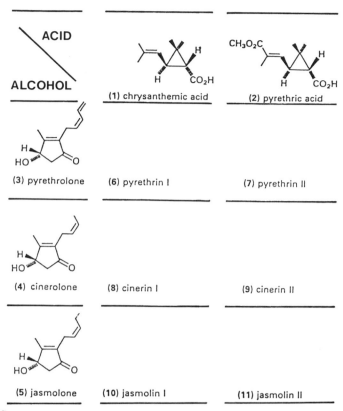

ACID / ALCOHOL	(1) chrysanthemic acid	(2) pyrethric acid
(3) pyrethrolone	(6) pyrethrin I	(7) pyrethrin II
(4) cinerolone	(8) cinerin I	(9) cinerin II
(5) jasmolone	(10) jasmolin I	(11) jasmolin II

Fig. 1   Structures and absolute configurations of the constituent acids and alcohols of the natural pyrethrins. The names of the six esters, known collectively as pyrethrins, are given. (Modified from Davies, 1985.)

synthetic pyrethroids much more rapidly (5–10 times) than their *cis* counterparts, and shows some similarities to malathion carboxyesterases and *p*-nitrophenyl acetate carboxyesterase. Ruzo *et al.* (1978), in a detailed study of decamethrin metabolism in rats, have demonstrated that the insecticide and various metabolites derived from its acid and alcohol fragments are almost completely eliminated from the body in 2–4 days.

1.1.2   Stereochemistry

Comprehensive accounts of pyrethroid stereochemistry are to be found elsewhere (Tessier, 1985; Davies, 1985). Here it suffices to provide a brief explanation of the terminology most frequently encountered in the extensive literature on these compounds. The chemical groups attached to a saturated carbon atom can be arranged in two non-superimposable ways. They represent mirror images (enantiomers). Compounds with two (or n) asymmetric carbon atoms will have respectively 2^2 (and 2^n) possible isomers. If all the asymmetric centres in one isomer are mirror images of the corresponding centres on the other, then the isomers are mirror images or enantiomers. If a mixture of identical and non-identical configurations is present, the compounds are referred to as diastereomers. Enantiomers share the same physical properties (e.g. solubility, melting point). Cahn–Ingold–Prelog (CIP) rules (Cahn et al., 1956) provide an unambiguous method for describing such compounds. The four groups are arranged in an order of priority based on atomic number and the carbon atom is always viewed from the side opposite to that containing the lowest priority group.

Deltamethrin is thus referred to as (S)-α-cyano-3-phenoxybenzyl (1R, 3R)-3-(2,2-dibromovinyl)-2,2-dimethylcyclopropanecarboxylate, and is the single most active isomer of the ($2^3 = 8$) possible. The cyclopropane ring is numbered as illustrated (Fig. 2), and configurations 1R, 3R and alpha 5 are shown. With bromine atoms replaced by methyl groups, the description of C-3 stereochemistry changes to S because the priority of the C-3 substituent is altered. A simplified scheme, which is less confusing, is normally used to describe cyclopropane carboxylates (Elliott et al., 1974). By defining C-3 stereochemistry relative to that of C-1, which is defined in absolute terms, deltamethrin is written as (1R)-cis-3(2,2-dibromovinyl)-2,2-dimethylcyclopropanecarboxylic acid, and whatever group replaces the dibromovinyl group at C-3, the stereochemical description is unchanged.

1.1.3   Structure–activity relations

Pyrethroid structure–activity relations have been comprehensively reviewed (Elliott and Janes, 1977, 1978b; Davis, 1985) and here we simply summarize some of the key findings by way of introduction to this extensive and complex literature. A number of structural features are essential for insecticidal activity. If any one of these is not present there will be a significant drop in insecticidal activity of the compound.

(1) The vast majority of pyrethroids are esters of carboxylic acids;
(2) Active esters of 3-substituted cyclopropane carboxylic acids all take the 1R-configuration;
(3) A gem-dimethyl substituent at the C-3 position on the cyclopropane ring is normally required for activity;
(4) Derivatives of alkyl or aralkyl esters are the only ones that show activity. The carbon to which the hydroxy group is linked must be sp3

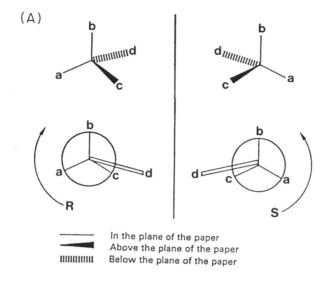

(A)

In the plane of the paper
Above the plane of the paper
Below the plane of the paper

(B)

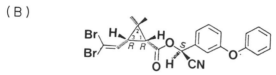

Fig. 2   (A) Description of absolute stereochemistry about an asymmetric carbon atom. The order of priority is a > b > c > d. (B) Absolute stereochemistry of the pyrethroid insecticide deltamethrin, (S-α-cyano-3-phenoxybenzyl) (1R, 3R)-*cis*-3-(2,2 dibromovinyl)-2,2-dimethylcyclopropanecarboxylate. (From Davies, 1985.)

hybridized (tetratredral). Phenyl esters do not possess insecticidal activity;

(5) Replacing the α-carbon atom of 3-phenoxybenzyl alcohol by a cyano group increases the activity of the esters to most insect species.

Most of the features listed above do not depend on the insect species under test and appear to relate to the intrinsic activity of the molecules. Although a wide range of structural features beyond those listed above have been systematically changed, the effects on insecticidal activity have been variable and do not permit general conclusions. However, the configuration at C-3 does have an influence on the susceptibility to metabolic detoxification. Studies with synergists (Casida, 1983) indicate that modifications at this site have little effect on intrinsic activity. Investigations by Soderlund (1983) have provided evidence that ease of access to the active site is influenced by the C-3 configuration.

1.1.4 *Selective toxicity of pyrethroids*

Pyrethroids show considerable selectivity for insects compared to mammals, with LD_{50} selectivity ratios (rat oral compared to insect topical application) of 4500:1 (Elliott, 1977). Mammalian neurones, unlike insect neurones, are insensitive to 1R, *trans* isomers of resmethrin and phenothrin. As discussed in subsequent sections, it may be that the introduction of an α-cyano substituent changes either the selectivity and/or the type and site of pyrethroid action. There is growing evidence that resistant insects with the kdr factor have modified target sites, as evidenced by a change in neuronal sensitivity to pyrethroids (Sawicki, 1985). However, the "natural" tolerance to pyrethroids displayed by some insect species appears to be related to enhanced detoxification.

The major contribution of metabolism to the selective toxicity of pyrethroids has been clearly summarized by Casida *et al.* (1983). Species differences in rates of detoxification *in vivo*, activities of detoxifying enzymes *in vitro* and the response to detoxification inhibitors as synergists, all provide evidence for this view. The properties of purified pyrethroid carboxyesterase from rat liver microsomes have been well documented (Suzuki and Miyamoto, 1978). Selectivity is mainly due to differences in rates of detoxification rather than to differences in the sites of metabolic attack. However, higher levels of oxidase and esterase enzymes in resistant insect strains have been postulated to contribute to pyrethroid resistance in *Musca domestica* (Tsukamoto and Casida, 1967; Plapp and Casida, 1969). The tolerance of green lacewing larvae to pyrethroids appears to be attributable, at least in part, to an extremely high rate of esterase detoxification (Ishaaya and Casida, 1981). Synergism studies by Soderlund and Casida (1977) have shown that metabolism is a key factor in the low toxicity to mice of fenvalerate. Pyrethroid hydrolysing esterases have been found in the gut and integument of *Trichoplusia* larvae, and in the cuticle, gut and fat body of *Spodoptera* larvae (Ishaaya and Casida, 1980; Abdel-Aal and Soderlund, 1980). Differences in the capacity to metabolize pyrethroids may be one of a number of factors contributing to differences in susceptibility to pyrethroids between various insect species.

1.2 INSECTICIDAL ACTIONS

1.2.1 *Knockdown and lethal actions*

Fast-acting pyrethroids are designated good knockdown agents. Knockdown by a pyrethroid is the characteristic immobilization of the insect, and although it often precedes a lethal action, this is not always the case. Knockdown may result from actions on the central nervous system (CNS) of insects (Burt and Goodchild, 1974). Alternatively, it may arise from a combination of central and peripheral actions (Clements and May, 1977; Glickman and Casida, 1982). The precise mechanism remains to be fully resolved, but an early observation by Hutzel (1942) showed that pyrethrum poisoning progressed regardless of whether or not the cockroach ventral nerve cord was attached. More recently, Gammon (1977, 1978) showed that the cercal afferent, giant interneurone system of the cockroach was functional during all the early stages in which symptoms of knockdown were evident. It therefore appears that certain central neurones are still functional, even during the advanced stages of pyrethroid poisoning.

Briggs et al. (1974) and Ford (1979) showed that optimal polarity for knockdown by a series of pyrethroids was greater than the optimum polarity for lethal action. Nishimura et al. (1982) reached a similar conclusion. Clements and May (1977) suggested that structural features as well as this physical property are important in knockdown.

1.2.2   Pyrethroids as neurotoxic agents

Both central and peripheral actions of pyrethroids have been demonstrated on target and non-target organisms. Miller and Salgado (1985) review much of the earlier work on this topic. Pyrethroid-poisoned mice and rats die during seizures within 1–2 hours of exposure (Ray and Cremer, 1979; Lawrence and Casida, 1982). Although a detailed account of the extensive toxicity studies on mammals and other non-target organisms is beyond the scope of this review, it is of interest to note the variety of effects of pyrethroids reported for the vertebrate central nervous system. Striking increases in activity in electroencephalogram records and extracellular recordings are noted throughout the central nervous system (Cremer, 1983) of pyrethroid-poisoned rats. Studies on the deltamethrin-induced motor syndrome in rats indicate that this effect is extrapyramidal in origin (Brodie and Aldridge, 1982; Brodie, 1983) and is accompanied by elevated cerebellar cyclic GMP. The motor syndrome can be reversed by mephensin (Bradbury et al., 1983). Reduction of acetylcholine (ACh) content of brain has also been reported (Aldridge et al., 1978) along with increased glucose utilization, particularly in the cerebellum and hypothalamus (Cremer et al., 1980, 1981), and changes in cerebral blood flow (Cremer and Seville, 1982; Ray, 1982) accompany pyrethroid poisoning.

It may be that insects are less dependent than mammals on continuous nervous control of respiration and circulation. As a result, the point of death

TABLE 1   Insecticidal activities and neurotoxicity of pyrethroids

Compound	Relative potencies (Bioresmethrin = 100)		Knock down	Neurotoxicity (Bioresmethrin = 100)
	Musca domestica	*Periplaneta americana*	*Musca domestica*	*Periplaneta americana*
Bioresmethrin	100[a]	100[b]	—	100[c]
1S enantiomer	0·8	0·4	—	7
Biopermethrin	74	580	—	45
1S enantiomer	0·7	—	—	33
Biotetramethrin	0·4	6	+	52
S-bioallethrin	9	300	+	131
Cismethrin	42	400	—	149
Deltamethrin	2800	3000	—	3
Kadethrin	39	80	+ +	279

[a] LD_{50} 0·005 g/insect.
[b] LD_{50} 2·5 g/insect.
[c] Relative potency based on pyrethroid-induced after-potentials in cockroach axons after 16 min exposure to compound.
Data from Elliott and Janes (1978a); Elliott *et al.* (1978); Burt *et al.* (1974); P. E. Burt (personal communication); Laufer *et al.* (1984). Table from Laufer *et al.* (1984).

is not readily defined. It may well result from irreversible damage to the nervous system when poisoning has proceeded for several hours. Adams and Miller (1979a, b) concluded on the basis of flight motor discharge (extracellular) recording that despite the sensitivity to tetramethrin, the role of motor unit discharge in the events leading to death is dubious. This is because tetramethrin-induced discharges disappear at lower temperatures, whereas toxicity increases.

Recovery from pyrethroid poisoning was studied in houseflies treated with either *trans*-permethrin or deltamethrin (Bloomquist and Miller, 1986b). The first sign of recovery was the appearance of normal posture, which was followed by jumping behaviour and finally, coordinated flight. Prior to full recovery, treated houseflies were able to maintain normal posture and could usually jump, but they were unable to fly. When tethered, these flightless houseflies responded to loss of tarsal contact by initiating normal patterned activity in the dorsolongitudinal flight muscles, yet the wings did not move. In flightless flies displaying jumping behaviour, electrical stimulation of the brain evoked responses in the pleurosternal muscle, which controls thoracic tension during flight. Extending studies of this type to the single cell level may help in understanding which elements of the nervous system are affected most profoundly by pyrethroid poisoning.

1.2.3   Type I and Type II pyrethroids

Based on the symptoms of pyrethroid action on whole insects, two distinct classes of pyrethroids (Type I and Type II) have been postulated. More recently, this view has been extended to incorporate two distinct sites of action of the subclasses of pyrethroid insecticides responsible for these effects.

Gammon *et al.* (1981) subdivided pyrethroids into 2 classes (Types I and II) based on their effects on cercal sensory neurones recorded *in vivo* and *in vitro*, and on the symptomology they produced in dosed cockroaches (*Periplaneta americana*). Type I compounds included pyrethrins, S-bioallethrin (1R, *cis*) resmethrin, kadethrin, the 1R, (1) *trans* and 1R, (1) *cis* isomers of tetramethrin, phenothrin and permethrin, and an oxime O-phenoxybenzyl ether. Narahashi (1980) proposed the following terminology: Type I for the pyrethroids which induce repetitive firing; Type II for those pyrethroids which depolarize the resting membrane.

Electrophysiological recordings from treated cockroaches revealed increased frequency of cercal sensory nerve action potentials and sometimes also action potentials from the cercal motor fibres and unidentified neurones in the CNS (Gammon *et al.*, 1981). Low concentrations of these compounds act *in vitro* to induce repetitive firing in a cercal sensory nerve following a single electrical stimulus. Based on these *in vitro* measurements, it was concluded that only the 1R insecticidal isomers were highly effective neurotoxins. The most potent compounds on the isolated nerve were (1R, *trans*)- and (1R, *cis*) tetramethrin. The poisoning symptoms of Type I compounds are restlessness, incoordination, hyperactivity, prostration and paralysis.

Type II compounds include (1R, *cis* α-S)- and (1R, *trans* α-S) cypermethrin, deltamethrin and (S,S) fenvalerate. These α-cyanophenoxybenzyl pyrethroids, did not induce repetitive firing either *in vivo* or *in vitro*; they caused different symptoms, including a pronounced convulsive phase. Two other pyrethroids with an α-cyano substituent, i.e. fenpropathrin and an oxime 0-α-cyanophenoxybenzyl ether are classified as Type I, based on their actions on cockroach cercal sensory nerves, but the symptoms observed with these compounds resemble those normally associated with Type II compounds.

Some authors regard these categories as separating primarily central nervous system actions (Type II), as exemplified by α-cyano-containing pyrethroids, from peripheral actions by other (Type I) pyrethroids such as allethrin and bioresmethrin (Gammon *et al.*, 1981; Staatz *et al.*, 1982). Separate categories of action of pyrethroids can also be constructed by examination of the pattern of repetitive discharges in flight motor nerves during pyrethroid poisoning (Adams and Miller, 1980). Nevertheless, there are exceptions to

these distinct categories of action (Gammon *et al.*, 1981; Scott and Matsumura, 1983).

1.2.4   *Temperature sensitivity*

Blum and Kearns (1956) demonstrated that pyrethrum was more effective on
Periplaneta americana at temperatures lower than normal ambient temperature. This negative temperature coefficient is present to varying degrees
depending on the pyrethroid and the insect species under investigation. Temperature dependence of nervous activity has been examined by several laboratories. Narahashi (1971) reported that allethrin induced repetitive
discharges in cockroach nerve cords only below 26 C. This negative temperature coefficient also extended the actions of allethrin on isolated squid giant
axons (Narahashi, 1971). Romey *et al.* (1980) showed a break in Arrhenius
plots of membrane conductance between 17 C and 20 C. This, together with
differences in the actions of veratrine above and below 20 C, suggested that
the fluidity of membrane lipids in the vicinity of the sodium channel might
affect the properties of the channel gating mechanism. Since binding of the
specific sodium channel blocker [^{3}H]tetrodotoxin did not vary over the temperature range 1–30°C in purified crab axonal membrane, clearly channel
density remained the same. Hendy and Djamgoz (1985) propose that deltamethrin-induced accumulation of extracellular potassium in cockroach nerve
cords, following low-temperature block of the neuronal sodium/potassium
exchange pump, may explain the negative temperature coefficient, but this
could not account for its detection in isolated, single axons (Narahashi, 1971)
and cultured neurones (Chinn and Narahashi, 1987). Further studies are
needed on the molecular basis of this intriguing phenomenon.

1.2.5   *Resistance*

The development of resistance to pyrethroid insecticides threatens the long-
term usefulness of the compounds for the control of arthropod pests. A comprehensive review of pyrethroid resistance by Sawicki (1985) provides an excellent overview of this problem. Resistance includes any tolerance to a
pyrethroid greater than that of the most sensitive reference (= susceptible)
strain. It can develop under field conditions due to excessive and continued
application. Laboratory selection and cross-resistance studies can also lead
to the appearance of resistance. Where pyrethroids have been used in field
conditions over many years they have probably reselected cross-resistance
mechanisms that first appeared in response to an earlier generation of insecticides. A well-characterized mechanism responsible for cross-resistance
between the organochlorine insecticide DDT and pyrethroids is known as

kdr. In the housefly *Musca domestica* it delays the onset of knockdown by DDT and pyrethroids. Early work by Milani (1954) and Farnham (1977) isolated genetically the *kdr* factor and showed it to be controlled by a recessive gene on autosome 3. It is linked to green eye and brown body genetic markers in a well defined housefly strain (538ge), which proved to be cross-resistant to all pyrethroids tested and to DDT (Sawicki, 1978).

It appears that a single gene (or locus) is responsible for conferring site insensitivity (Farnham, 1977). In the context of understanding the mechanism of pyrethroid resistance studies on *kdr* flies are proving of considerable value, even though the *kdr* gene product is not known. For example, presynaptic terminals (Salgado *et al.*, 1983a) and axons (Gammon, 1980; Salgado *et al.*, 1983b; Osborne and Smallcombe, 1983) of *kdr* insects were less sensitive to pyrethroids than their susceptible counterparts. This may be due to a change in a membrane protein molecular target (De Vries and Georghiou, 1981a, b; Miller *et al.*, 1979). However, it should be recalled that Chialiang and Devonshire (1982) demonstrated changes in the lipid properties of housefly head membranes using Arrhenius plots of the activity of membrane-bound acetylcholinesterase. Recently, Fukami and colleagues working on resistance in the 228e2b strain of housefly employed a combination of linkage group analysis and electrophysiology to show that the recessive genetic factor responsible for nerve insensitivity is located on housefly chromosome 3 (Ahn *et al.*, 1986a, b). Further studies with modern methods of electrophysiology, neurochemistry and molecular biology offer experimental approaches to understanding this phenomenon, which is of considerable practical significance for the future of pyrethroids.

2 Pyrethroid–channel interactions

2.1 SODIUM CHANNELS

Sodium channels are well known membrane proteins which produce the rising phase of action potentials in nerve, muscle, and many other excitable tissues. In the strict sense that we adopt in this review, the sodium channels are characterized by their unique ion selectivity to Na^+ and Li^+ over other cations, their voltage-regulated gating mechanism, and their high sensitivity to tetrodotoxin (TTX) (Hodgkin and Huxley, 1952; Narahashi, 1974; Hille, 1984; Yamamoto, 1985). By this definition, therefore, many membrane channels are not regarded as sodium channels, even if they are highly permeable to sodium ions (e.g. Benos *et al.*, 1986). Indeed, the sodium channels are a family of homologous membrane proteins with relative molecular mass

$\simeq 260,000$ (Catterall, 1986). The amino acid sequences are highly conserved not only for different types of sodium channels in the same organism (Noda *et al.*, 1986), but also between the equivalent molecules in different animal species (Noda *et al.*, 1984; Salkoff *et al.*, 1987). Sodium channel neurotoxins are proving useful in probing the knockdown resistance (*kdr*) mechanism in houseflies (Bloomquist and Miller, 1986a). It is almost certain that the primary site of action of pyrethroids is the sodium channel, and contributory evidence is discussed in the remainder of this section.

2.1.1 *Extracellularly recorded neuronal activities of pyrethroids*

Effects of pyrethroids on excitable cells are detected most easily by monitoring impulse discharges from nerves and muscles with extracellular recording electrodes. Some of the most obvious changes revealed by this approach are the elevated frequency of spontaneous firing and the prominent after-discharges following single stimulatory inputs to the nervous system. For example, in axons of *Rhodnius prolixus* neurosecretory cells, the discharge rate jumped to 150 Hz from the control value of 40 Hz within 5 minutes of application of 1.0×10^{-8} M bioresmethrin (Orchard, 1980). A similar result (see Fig. 3a) was obtained in neurosecretory axons of the stick insect *Carausius morosus* (Orchard and Osborne, 1979). In the cercal afferent, giant interneurone preparation of the cockroach *Periplaneta americana*, a single electrical shock applied to the presynaptic cercal nerve evokes a large compound action potential, which lasts for less than 5 ms under normal conditions. After exposure of the preparation to nanomolar concentrations of allethrin, permethrin or cypermethrin, the compound action potential was invariably accompanied by a long train of spikes lasting for several seconds (Gammon, 1978, 1979, 1980; Gammon *et al.*, 1981, 1983). The propagation of these spikes in the interneurone giant axons is orthodromic in the sense that the direction of conduction along the axons is ascending, as is also the case for control preparations.

Adams and Miller (1979a) found in the flight motor system of *Musca domestica* that tetramethrin stimulated antidromic discharges in addition to orthodromic spike activities (Fig. 3b). By recording the motor activity at two separate points along the nerve, they revealed that a single orthodromic nerve spike, upon reaching the motor axon terminals, triggered a discharge which passed postsynaptically and also backfired into the proximal part of the axon (Adams and Miller, 1979a, b). It was suggested that the repetitive discharges originated from a restricted region in the terminal intramuscular portions of motor nerve fibres (Adams and Miller, 1979a). However, these authors also pointed out that the role of the backfiring in the events leading to death of the insect was questionable, since motor unit discharges

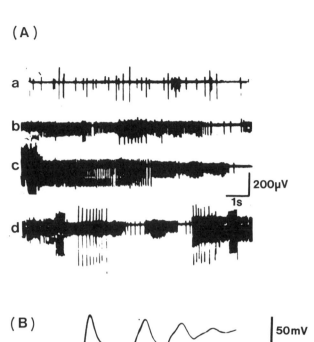

Fig. 3   (A) The effects of permethrin on the spontaneous activity recorded from the isolated transverse nerve in *Carausius morosus*. (a) Control recording; (b, c, d) records obtained 24, 30, and 90 min, respectively, after the addition of 5.0×10^{-8} M permethrin, Scale bars: 200 μV, 1 s. (B) Repetitive discharges recorded along the posterior dorsal median nerve (lower two traces) and from dorsal longitudinal muscle DLM5 (upper trace) of the housefly, *Musca domestica*. The first muscle potential in the repetitive discharge is shown to be driven by a normal, orthodromic nerve impulse emerging from the thoracic ganglion. The lower trace shows the nerve spike as it passes the most proximal electrode; the same impulse is recorded 0·5 ms later in the distal electrode (middle trace), followed by the muscle intracellular potential (upper trace). Subsequent muscle potentials are correlated with antidromic nerve impulses, evidenced by the arrival of a nerve spike first at the distal electrode, then at the proximal electrode. Calibration: 50 mV (upper trace), 100 mV (middle and lower trace); 5 ms (all traces). [(A) From Orchard and Osborne (1979); (B) from Adams and Miller (1979).]

disappeared at lower temperatures, whereas toxicity increased (Adams and Miller, 1979a, b). The repetitive firing at the presynaptic motor terminals resulted in synaptic malfunction. In the presence of 100 nM cismethrin, neuromuscular transmission was blocked following a massive release of transmitter from the presynaptic terminal, as detected by the drastic increase in frequency of miniature excitatory postsynaptic potentials (mEPSPs) in the muscle (Salgado et al., 1983a, b). The acceleration of transmitter release was much more evident for deltamethrin, a Type II pyrethroid. Block was primarily due to the exhaustion of synaptic vesicles in the terminal (Schouest et al., 1986).

However, in the later stage of neuromuscular block, responsiveness of the postsynaptic membrane to the candidate transmitter, L-glutamate, was also impared by deltamethrin (Salgado et al., 1983b). Although the mechanism of this is unknown, the authors consider that block of the glutamate response by pyrethroids may be related to a presynaptic action rather than a separate postsynaptic action, because treatment of the preparation with lanthanum and respiratory inhibitors, presumably acting presynaptically, also had the same effect. It is possible that many of the postsynaptic receptors were internalized or down-regulated after prolonged exposure to the transmitter. Such changes may correspond to the desensitization of β-adrenergic receptors observed following treatment with high doses of adrenaline (Sibley and Lefkowitz, 1985).

Actions of pyrethroids on sensory receptor organs of insects are also well documented. For instance, repetitive discharges in the crural nerve of the locust seen after application of a Type I pyrethroid ceased immediately after removal of the chordotonal organ, indicating that most of the abnormal nervous activity recorded from this nerve arises from that organ (Clements and May, 1977).

High sensitivity of sensory nerves to pyrethroids has also been suggested by a behavioural experiment, in which the flight reflex is tested by examining whether or not tethered flies (*Musca domestica*) initiated flight upon removal of the substrate (Bloomquist and Miller, 1985). Blockage of the flight reflex by permethrin occurred approximately 13 minutes after treatment. As carbofuran is known to initiate convulsive flight in houseflies, it was used as a chemical probe of the integrity of the flight system in the central nervous system. Topical application of carbofuran ($1 \cdot 0 \, \mu$g) to tethered flight reflex preparations already blocked by permethrin elicited spontaneous flight. On the basis of these results, Bloomquist and Miller (1985) suggested that pyrethroid block of the flight reflex was due to an action on sensory nerves, since the central flight programme and its associated efferent systems were functionally intact. In contrast to the actions of Type I pyrethroids, Type II pyrethroids such as deltamethrin never induced bursts of activity in a mechanoreceptor cell of the cockroach (Guillet *et al.*, 1986). Deltamethrin had little effect on the receptor potential induced by mechanical stimulation of the sensory hair but blocked the action potentials within a few minutes (Guillet *et al.*, 1986).

The results discussed so far were all from experiments with the target organisms, i.e. insects. Actions of pyrethroids on the vertebrate nervous system are very much in line with those on insects (Wouters and van den Bercken, 1978). Staatz-Benson and Hosko (1986) have reported an exceptional case. They found that not only *cis*-permethrin but also deltamethrin had dramatic facilitatory effects on spontaneous firing rates of ventral roots and spinal interneurones in the rat. Generally speaking, Type II pyrethroids such as deltamethrin cause conduction block without provoking repetitive firing in other systems.

Although pyrethroids exert a variety of behavioural and physiological effects as mentioned above, most of these appear to be mediated via an action on the sodium channel, as they are blocked by a selective sodium channel blocker, TTX in target as well as non-target organisms (Berlin *et al.*, 1984; Ghiasuddin and Soderlund, 1985; Roche *et al.*, 1985; but see Leake *et al.*, 1985).

2.1.2 *Intracellular recordings with conventional microelectrodes*

Although extracellular recordings of nervous activity successfully reveal neurophysiological effects of pyrethroids, they are not very helpful in dissecting possible mechanisms of action. Intracellular recordings with conventional microelectrodes can provide insight into mechanisms whereby pyrethroids produce repetitive discharges and/or block of conduction in nerves and sense organs.

In the crayfish stretch receptor organ, bursting activity associated with pyrethroid-treatment (permethrin and bioallethrin) seems to be initiated by a depolarization of the resting membrane potential (Chalmers and Osborne, 1986a). The α-cyano substituted pyrethroids (cypermethrin and deltamethrin) produced gradual depolarization of the membrane without inducing repetitive firing, resulting in an eventual block of action potential generation (Chalmers and Osborne, 1986a). These actions are blocked by TTX, indicating involvement of the sodium channel (Chalmers and Osborne, 1986b). Ruigt *et al.* (1986) have reported that during repetitive activity of stretch receptor cells, the soma membrane gradually depolarized and the action potential decreased in amplitude in the presence of fenfluthrin. The falling phase of the action potential became prolonged and the hyperpolarizing afterpotential was gradually displaced in the depolarizing direction. They also noted that in the presence of $0.5 \mu M$ TTX the shape, amplitude and stretch-dependence of the generator potential were unaffected by fenfluthrin (Ruigt *et al.*, 1986). Using the voltage-clamped crayfish axon, Lund and Narahashi (1983) have shown very clearly that repetitive firing in the presence of the Type I pyrethroids is produced by the depolarizing afterpotential which is absent in control conditions (Fig. 4). In marked contrast to the Type I compounds, Type II pyrethroids cause little or no depolarizing afterpotential or repetitive firing in response to a single stimulus, but they depolarize the axon to the extent that the action potential is blocked (Lund and Narahashi, 1983).

Thus the mechanisms responsible for generating the depolarizing afterpotentials and the gradual depolarization of the resting potential are the most important issue to resolve in understanding the mode of action of pyrethroids. The voltage-clamp technique is the most appropriate experimental approach for this type of study.

2.1.3 *Voltage-clamp analysis*

Cockroach giant axons are amenable to voltage-clamp experiments using the oil-gap, single-fibre recording method (see Pelhate and Sattelle, 1982). By depolarizing the membrane from a holding potential of $-60\,\text{mV}$ to potentials between $-30\,\text{mV}$ and $+50\,\text{mV}$ with command voltage steps of 5 ms duration, a single component of transient sodium current can be activated following potassium channel block by 3,4 diamino-pyridine. Tail currents upon repolarization are not detectable in this condition. In axons treated with Type I pyrethroids, the transient current was accompanied by a noninactivating sodium component. A huge inward tail current lasted for 1 s following repolarization of the membrane to $-60\,\text{mV}$ (Laufer *et al.*, 1984, 1985). The holding current was minimally affected by the Type I pyrethroid. In contrast, deltamethrin, a Type II pyrethroid, induced a large increase in the holding current at $-60\,\text{mV}$ (Laufer *et al.*, 1985). The actions of deltamethrin on cockroach (*Periplaneta americana*) axonal sodium currents are illustrated in Figure 5.

Quantitative analyses of the actions of pyrethroids on gating kinetics of sodium channels were performed using data obtained from giant axons of crayfish (Lund and Narahashi, 1981a) and squid (Lund and Narahashi, 1981b) by means of sucrose-gap and axial-wire voltage-clamp techniques, respectively. Conclusions drawn from experiments using these different preparations were practically the same. Many of the experiments with voltage-clamp and patch-clamp (see Section 2.1.4) employ concentrations in excess of those used for detecting pyrethroid actions with extracellular hook electrodes (see Section 2.1.1) or even microelectrodes (2.1.2). This is largely for practical constraints imposed by the difficulty of clamping membranes for extended

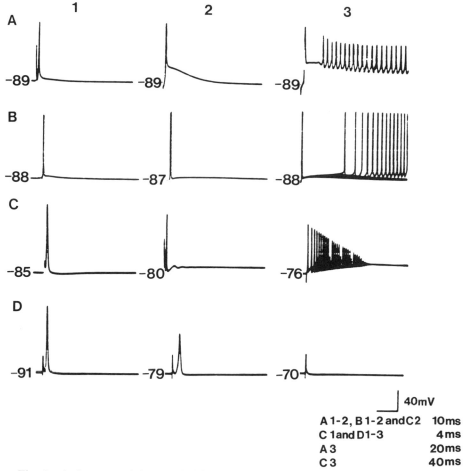

Fig. 4   Action potentials recorded from crayfish giant axons before (column 1) and after (colums 2 and 3) treatment with pyrethroids and DDT: 1.0×10^{-5} M DDT (8 min for A2 and 21 min for A3), 3.0×10^{-8} M tetramethrin (14 min for B2 and 23 min for B3), 1.0×10^{-6} M phenothrin (9 min for C2 and 18 min for C3), and 1.0×10^{-7} M cyphenothrin (5 min for D2 and 15 min for D3). Time calibration: For A1–2, B1–2, and C2, 10 ms; for C1 and D1–3, 4 ms; for A3, 20 ms; for C3, 40 ms. Potential calibration: 40 mV. The numbers represent the resting potential (mV) immediately preceding the action potential. (From Lund and Narahashi, 1983.)

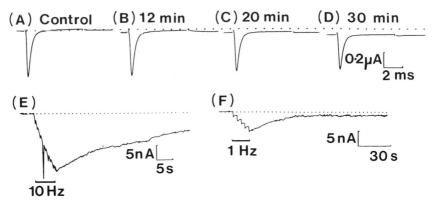

Fig. 5   Actions of deltamethrin on cockroach axonal sodium currents: (A) control in normal saline shows the sodium currents recorded at a holding potential of $-60\,mV$ (dashed line indicates zero current) and the peak inward sodium current associated with a 5 ms voltage-clamp pulse to $E_m = -10\,mV$; (B, C and D) after application of deltamethrin for, respectively, 12, 20 and 30 min, a reduction in the amplitude of the peak inward sodium current was observed, together with the progressive appearance of a maintained inward sodium current; in (D), this current was 80 nA, corresponding to 12·5% of the peak inward current recorded in control experiments; (E and F) slow sweep-speed, high-gain recordings of sodium currents; the peak inward (transient) current was not recorded because of the low value of the digitizing frequency; the currents were recorded after bursts of voltage pulses (5 ms duration, to $E_m = -10\,mV$) at the frequencies indicated; dashed lines correspond to inward sodium currents of 50 nA in (E) and 58 nA in (F); recordings were obtained 18 min (E) and 24 min (F) after application of $2\,\mu M$ deltamethrin; the slow (and incomplete) decline of these burst-induced inward sodium currents can be seen; holding potential $(E_h) = -60\,mV$. (From Laufer *et al.*, 1985.)

periods and all the evidence to date indicates that the rapid effects at higher concentrations are similar to those that develop more slowly following extended applications at lower concentrations.

In the presence of $3{\cdot}0 \times 10^{-5}\,M$ tetramethrin, the early transient sodium current of the squid axon was followed by a sustained component with a slowly-decaying tail current (Fig. 6), as in the case of cockroach axons. The early sodium current seen in tetramethrin-treated axons retains the characteristics of this component in normal axons, in that the time-courses of both the rising phase (activation, *m*) and the falling phase (inactivation, *h*) are indistinguishable from those of untreated membranes over a wide range of membrane potentials (-60 to $-30\,mV$). The fastest of the three components of the tail current seen in tetramethrin-treated axons decays at a similar rate to the control tail current, indicating that tetramethrin has not influenced the inactivation process of the early sodium current. These results led Lund and Narahashi (1981a, b) to suggest that the early component of the sodium

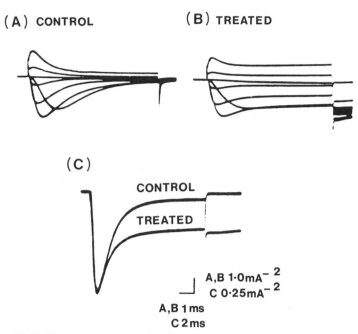

Fig. 6 (A) Sodium currents associated with step depolarizations from the holding potential of -100 mV to -50, -30, -10, 0, $+10$, $+30$, $+50$ and $+70$ mV in a control squid axon. (B) Sodium currents recorded as in (A) but after internal treatment with 3.0×10^{-5} M tetramethrin. (C), sodium currents recorded at -10 mV before and 6 min after treatment with tetramethrin. Scale: 1 mA.cm^{-2} and 1 ms for (A) and (B); 0.25 mA.cm^{-2} and 2 ms for (C). (From Lund and Narahashi, 1981a.)

current observed in the treated axon represented a fraction of the sodium channels that remained unaffected even in the presence of tetramethrin.

The two slower components of the tail currents of tetramethrin-treated axons are not seen in control conditions. The decay rates for these tail cur-rents decrease with depolarization. The time-constant for the slowest com-ponent was about 60 ms at -80 mV, i.e. 300 times larger than the control value. The slow tail current is undoubtedly monitoring the fraction of sodium channels modified by tetramethrin. This finding indicates that tetra-methrin drastically retards conformational changes associated with the m-gate of the sodium channel.

It is quite clear that tetramethrin greatly slows down the inactivation of the squid axon sodium current, as the rate of decay of the current during the depolarizing step is very slow. The steady-state inactivation ($h\infty$) curve obtained from tetramethrin-treated axons by the two-pulse protocol (pre-pulse length, 160 ms) has a distinct base close to 0 mV, indicating the existence of incomplete inactivation at the end of 160 ms pulse sodium current, i.e. the current flowing through the modified-channel. The $h\infty$ curve for the inactivating current in the presence of tetramethrin was shifted in the negative direction by about 10 mV. Recovery from the inactivation was measured by applying two identical step depolarizations to the membrane at varying intervals. The recovery time-course followed second-order kinetics, with time-constants of the order of 10^{-3} s and 10^{-1} s in the absence and presence of tetramethrin. The slow component contributed only about 6% in control conditions, whereas it was almost 50% in axons treated with 1.0×10^{-5} M tetramethrin. Lund and Narahashi (1981b) think two equally plausible interpretations can account for this observation: (1) the channel dwells for a considerable time in a new drug-induced inactivated state; (2) slow inactivation, known to exist in normal channels, is greatly enhanced by tetramethrin.

The development of inactivation of sodium currents can be detected by measuring the instantaneous amplitude of tail currents upon repolarization from the conditioning command steps of fixed amplitude and varying duration. Experiments have shown that the amplitude of tail currents increases as the duration of the conditioning pulse increases up to 160 ms. With a further increase in duration of the command step, the tail current amplitude tends to decline. The initial rise in the tail current amplitude may be explained as slowing of the sodium current activation by the binding of tetramethrin to the channel in its resting state. The decrease in the tail current amplitude with longer conditioning pulses may result from the inactivation of the pyrethroid-modified sodium channel. A close inspection of the relationship between tail current amplitude and command pulse duration has revealed that the development of the tail current, which is a measure of the rate of arrival of channels in the modified open state, is a double exponential function, with time-constants of 1.3 ms (t_1) and 24.3 ms (t_2). Interestingly, removal of inactivation by treating the cytoplasmic face of the membrane with pronase or N-bromoacetamide eliminated the slow component. From this experiment, Lund and Narahashi (1981b) conclude that the second phase of the time-dependent tail current growth seen in the presence of sodium inactivation results from reopening of channels from the inactivated state to the modified open state.

Thus, the modified open state can be reached from at least two different states: (a) the modified resting state; and (b) the modified inactivated state. The normal open channel may also be modified, though evidence for this is equivocal. Lund and Narahashi (1981a, b) argue that this would happen because the fast phase of arrival of channels in the modified open state, as measured by tail-current amplitude, was greatly potentiated by removal of the competing pathway, i.e. the route to the normal inactivated state from the normal open state. Type II pyrethroids result in essentially the same

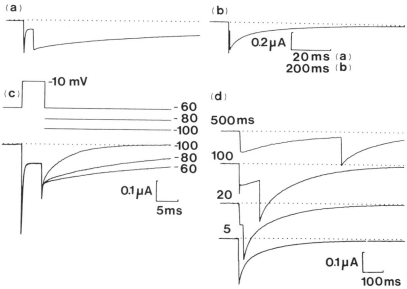

Fig. 7   Characteristics of the tail sodium current induced in a cockroach axon by methanotetramethrin: (a and b) sodium currents associated with 5 ms voltage pulses from − 60 mV to − 10 mV, followed by a return to − 60 mV; records were obtained in the presence of the K$^+$ channel blocker 3,4-diaminopyridine; the time-constant of the sodium tail current in such experiments was in the range 80–100 ms; (c) sodium currents (lower traces) associated with the voltage pulses shown in the upper traces; the time-constant of the tail sodium current decreased at more negative potentials; (d) sodium currents associated with voltage pulses from − 60 mV to − 10 mV, returning to − 60 mV, but of varying duration (between 500 ms and 5 ms); the tail sodium current time-constants were 80 ms after a pulse of 5 ms in duration, 125 ms after a 20 ms pulse, and 170 ms following application of a 500 ms pulse. (From Laufer et al., 1985.)

effects on the sodium channels as Type I compounds. Both the slow sodium current during step depolarization and the tail sodium current following step repolarization are smaller in amplitude when channels are modified by Type II as opposed to Type I pyrethroids, but the tail current decays much more slowly in the former case (Narahashi, 1985). Pyrethroid-induced tail sodium currents have also been detected in cockroach axons (Laufer et al., 1985), and examples are shown in Fig. 7 for methanotetramethrin-treated axons.

Very recently, Salgado and Narahashi (1988) examined the actions of fenvalerate (a Type II pyrethroid) on the gating current of the crayfish giant axon where the charge movement appears very large in comparison with squid axons (Swenson, 1983). They found that when sodium channels were driven into the modified-open state by repetitive stimulation in the presence of fenvalerate, the intermediate component of the ON gating current

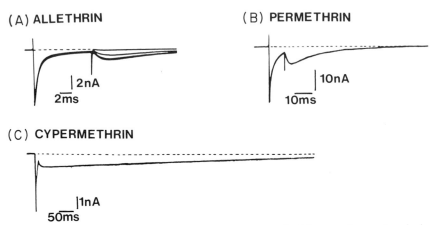

(A) ALLETHRIN   (B) PERMETHRIN

| 2nA
2ms

| 10nA
10ms

(C) CYPERMETHRIN

|1nA
50ms

Fig. 8   Sodium currents and sodium tail currents evoked by membrane depolarizations to $-5\,mV$, in frog myelinated nerve fibres treated with $2{\cdot}0 \times 10^{-5}\,M$ allethrin for 1, 2, 3 and 4 min (A) and steady-state effects of $1{\cdot}0 \times 10^{-5}\,M$ permethrin (B) and $1{\cdot}0 \times 10^{-6}\,M$ cypermethrin (C). (From Vijverberg and de Weille, 1985.)

$(t = 150\,\mu s$ at $+20\,mV)$, which corresponds to sodium channel activation, and the fast component of the OFF gating current $(t = 50\,\mu s$ at $-160\,mV)$, which corresponds to sodium channel deactivation, were inhibited in parallel with activation and deactivation of the sodium current. The slow component of the ON gating current, which has the same time-constant as sodium inactivation $(t = 600\,\mu s$ at $+20\,mV)$, as well as sodium current inactivation were abolished by fenvalerate. As the fast component of the ON gating current $(t = 45\,\mu s)$ was not affected by fenvalerate, Salgado and Narahashi (1988) have suggested that it is not associated with sodium channel gating. The immobilization of intermediate ON gating charges and fast OFF gating charges may imply that the activation gate is stuck in the open configuration. The immobilization of slow ON gating charges suggest that the inactivation gate is stuck in the resting position. Thus these observations on gating currents provide an elegant explanation of the appearance of sustained sodium current in pyrethroid-treated axons.

   A wide range of pyrethroids have been tested on the sodium current in frog nodes of Ranvier (Vijverberg et al., 1982, 1983). Unlike that of other preparations, the sodium current in pyrethroid-treated frog myelinated nerve fibres inactivated during step depolarizations, in much the same way as the normal axon does (Vijverberg and de Weille, 1985). However, on returning the membrane potential to the holding potential, a huge inward tail current preceded by a prominent rising phase appears (Fig. 8). Thus there is a clear difference between the frog and the other species that have been studied.

In contrast to the dramatic effects on the gating machinery of the sodium channel, pyrethroids have very little effect on the permeability and selectivity properties of the channel (Yamamoto *et al.*, 1986; Salgado and Narahashi, 1983).

2.1.4   *Patch-clamp analysis*

Perhaps the most direct experimental approach to investigating pyrethroid sodium channel interactions is to examine the effects of these compounds on the properties of single ion channels. This is now technically possible by means of the recently established patch-clamp method (Neher and Sakmann, 1976; Horn and Patlak, 1980; Hamill *et al.*, 1981). The gigaohm-seal patch-clamp technique also allows us to investigate macroscopic whole-cell currents of cells that have hitherto not proved amenable to voltage-clamp experiments.

The first successful study of pyrethroid actions at the single channel level was done by Yamamoto *et al.* (1983). These authors recorded sodium channel activity in inside-out as well as outside-out membrane patches isolated from mouse neuroblastoma N1E115 cells. Depolarization of the patch membrane to -50 mV or less negative potentials from a holding potential of -90 mV generated pulse-like inward currents due to the opening of sodium channels. The mean amplitude at -50 mV was approximately 1·5 pA, and the conductance calculated from the change in current size at different potentials was about 10 pS at 10 C. A Poisson plot of the open-times of the conducting states indicated a single exponential distribution. The mean rate-constant for channel closure was typically 0·588 ms^{-1}. In the presence of tetramethrin $(3\cdot0-6\cdot0 \times 10^{-5}$ M), the single channel conductance was not appreciably changed, whereas the distribution of open-times in the Poisson plot was better fitted by a sum of two exponential functions (Fig. 9). A portion of the population has a slower closing rate (0·059 ms^{-1}) than normal. The remaining exponential component had a rate constant of 0·526 ms^{-1}, which is close to that for the normal channels. Both components disappeared by external application of $3\cdot0 \times 10^{-6}$ M tetrodotoxin, indicating that both populations of open-states are associated with sodium channels. These open states probably represent those for separate unmodified and modified channels, respectively. The sodium channel modified by tetramethrin appears to inactivate slowly because the open-time decreases with large depolarizations. The reverse rate-constant for opening (deactivation) of the channel also seems to decrease because some of the channels remain in an open-state for tens of ms even after the membrane potential is returned to a holding potential of -100 mV (tail channels, see Yamamoto *et al.*, 1984, and Fig. 10A). Spontaneous openings of single sodium channels at the holding potential

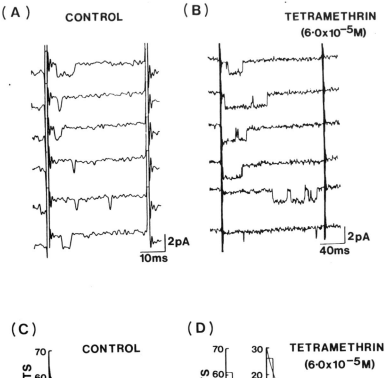

(A) CONTROL

(B) TETRAMETHRIN
(6·0×10⁻⁵M)

2pA
10ms

2pA
40ms

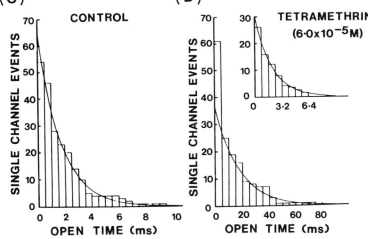

(C) CONTROL

(D) TETRAMETHRIN
(6·0×10⁻⁵M)

SINGLE CHANNEL EVENTS

OPEN TIME (ms)

Fig. 9 Effects of $6·0 \times 10^{-5}$ M (+)-*trans* tetramethrin on single sodium channels in an inside-out membrane patch excised from a neuroblastoma N1E115 cell. (A) sample records of sodium channel currents (inward deflections) associated with step-depolarizations from -90 mV to -50 mV. (B) as in (A), but after application of tetramethrin to the internal surface of the membrane. (C), channel open-time distribution in the control. (D), as in (C), but after application of tetramethrin. Inset shows the distribution of short open-times. (From Yamamoto *et al.*, 1983.)

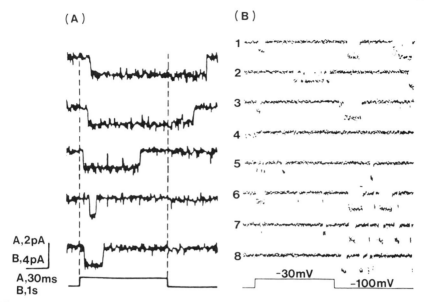

Fig. 10   (A) Single channel tail currents, and (B) post-pulse channel currents of NIE115 neuroblastoma cells recorded from inside-out membrane patches exposed to (A) 6.0×10^{-5} M tetramethrin and (B) 1.0×10^{-5} M deltamethrin. (A) The sodium currents were elicited by depolarizing the membrane to -40 mV from a holding potential of -100 mV. Two channels remained at the open-state even after stepping the membrane back to -100 mV. The amplitude of currents was almost identical during and after the depolarizing steps. (B) A series of consecutive current traces from another patch. Many channels opened after the membrane potential was stepped to the holding potential of 100 mV. [(A) From Yamamoto et al., 1984; (B) from Chinn and Narahashi, 1986.]

have never been seen in tetramethrin-treated patches. The patch-clamp observation is in keeping with the macroscopic data that tetramethrin does not increase the holding current in voltage-clamped giant axons (see Section 2.1.2).

Effects of deltamethrin on whole-cell and single sodium channel currents have been examined in the same NIE115 cells by Chinn and Narahashi (1986). The peak current–voltage (I–V) relationship of whole-cell sodium currents in deltamethrin-treated patches was different from the control curve in two respects. First, the modified channels could be activated at a potential 10–20 mV more negative than normal channels. Secondly, the I–V curve of modified channels rectified slightly between -10 and $+20$ mV. By measuring the tail current amplitude at a single potential on return from the different conditioning voltages, Chinn and Narahashi (1986) have demonstrated that the number of open channels diminished as the potential shifts from

-30 to $+10\,\mathrm{mV}$, and reaches a limiting value at more positive potentials, resulting in the appearance of rectification in the peak I–V relationship.

The effects of fenvalerate (Holloway *et al.*, 1984) and deltamethrin (Chinn and Narahashi, 1986) on the single sodium channel were broadly similar to those seen with tetramethrin, i.e. a prolongation of open-time without modifications of the unitary conductance of the channel.

Chinn and Narahashi (1986) made two further, interesting observations. First, they found that deltamethrin increased the number of blank traces (epochs with no channel openings) during repetitive step depolarization. The percentage of blank traces increased from 1% before drug application to 57% in traces examined between 20 and 40 minutes after continuous drug exposure. Secondly, they showed that the sodium channels modified by deltamethrin open not only during a depolarizing step, but also after repolarization (post-pulse channels, see Fig. 10B). In fact, this can also be detected with tetramethrin (D. Yamamoto, unpublished observations). The rising phase of tail currents observed in the pyrethroid-treated myelinated nerve fibre could be interpreted as the activation of post-pulse channels. The rising phase may be seen in the tail current of pyrethroid-treated squid axons, though it is not as clear as in the case of myelinated nerve fibres (Yamamoto *et al.*, 1986). Despite late openings after terminating a depolarizing step, no channels opened spontaneously without depolarizing steps in deltamethrin-treated patches (Chinn and Narahashi, 1986). These authors therefore conclude that deltamethrin stabilizes the various channel states by severely decreasing the rates of transition between them. They stress that this notion is also consistent with the observation of channel openings after a depolarizing pulse. For example, a channel that has closed after activation during the depolarizing pulse would be in a state in which the inactivation (h) gate was closed and the activation (m) gate was open (state A). If deltamethrin stabilized state A and after repolarization the h gate opened before the m gate closed, this would result in the channel opening after the pulse (Chinn and Narahashi, 1986). However, it is also possible that deltamethrin potentiated the slow inactivation, thereby increasing the blank traces during repetitive pulsing, as pointed out by Lund and Narahashi (1981a, b) in interpreting the slow recovery from inactivation of sodium currents in tetramethrin-treated squid axons. In fact, it has been implied that the microscopic counterpart to slow inactivation is the clustering of epochs with no channel openings (Horn *et al.*, 1984; Yamamoto, 1985).

The most recent result by Chinn and Narahashi (1987) has provoked even more complications. While looking at the effect of temperature on deltamethrin action, they found that at $11^\circ\mathrm{C}$ post-pulse channels had amplitudes similar to tail channels, whereas at $21^\circ\mathrm{C}$ two types of post-pulse channels were found: one of amplitude similar to that of tail channels, the other of much smaller amplitude—28% of tail channels (Chinn and Narahashi, 1987). On the other hand, Yamamoto *et al.* (1984) reported that the amplitude of the tail channels is much the same as those detected during depolarizing steps. The amplitude of the single channel current does not change much when the potential is stepped from $-30\,mV$ (pulse voltage) to $-100\,mV$ (holding voltage), because the expected increase in the amplitude due to the increased driving force is almost completely offset by the enhanced channel block due to the plugging effect of calcium ions (Yamamoto *et al.*, 1984).

Further work is needed to resolve these differences, which might be due to separate actions of pyrethroids on two populations of sodium channels. Distinct populations are known to exist in neuroblastoma cells (Nagy and Kiss, 1983; Yamamoto and Narahashi, 1983). That would also explain the conflicting observations reported for the batrachotoxin-modified single sodium channel. Quandt and Narahashi (1982) reported that batrachotoxin modification of the channel reduced unitary conductance by about 50%, while Huang *et al.* (1984) did not see any change in conductance. Alternatively, different open-states modified by pyrethroids might have different affinities for calcium ions, resulting in different channel amplitudes.

2.1.5 *Biochemical studies*

Biochemical approaches to the study of pyrethroid-sodium channel interactions have followed three courses. The binding of radiolabelled pyrethroid molecules to central nervous system (CNS) membrane extracts has been explored. In addition, unlabelled pyrethroids have been examined for their capacity to inhibit both the binding of specific sodium channel ligands, and the uptake of $^{22}Na^+$. Nevertheless, it was an important earlier biochemical study, eliminating a pharmacokinetic basis for differences in insecticidal activity of enantiomers, that led to the suggestion of a stereospecific membrane molecular target (Soderlund, 1979). This author showed that the potent pyrethroid insecticide NRDC 157 (3-phenoxybenzyl (1R,*cis*)-3-(2,2-dibromovinyl)-2,2-dimethylcyclopropane carboxylate) and its inactive 1S, *cis* enantiomer possessed similar cuticular penetration rates. They were detectable in the haemolymph, nerve cord, fat body, and midgut of the cockroach *Periplaneta americana*, within 2 hours of topical application at doses equivalent to the lowest lethal dose of NRDC 157. Both enantiomers showed a similar distribution pattern in the tissues. They were not detectably hydrolysed by fat body, nerve cord and haemolymph preparations *in vitro*. The findings provided no pharmacokinetic basis for the difference in insecticidal activity between enantiomers and indicated that the site of action of this pyrethroid was likely to be stereospecific.

Chang and Plapp (1983a, b) examined the binding of [^{14}C]-DDT and

PYRETHROID BINDING ASSAY

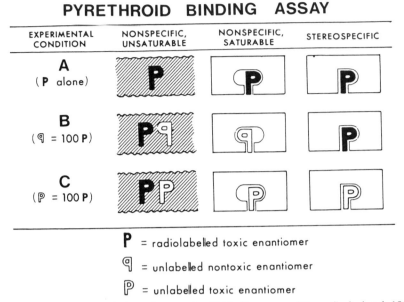

EXPERIMENTAL CONDITION	NONSPECIFIC, UNSATURABLE	NONSPECIFIC, SATURABLE	STEREOSPECIFIC
A (P alone)			
B (ꟼ = 100 P)			
C (ℙ = 100 P)			

P = radiolabelled toxic enantiomer

ꟼ = unlabelled nontoxic enantiomer

ℙ = unlabelled toxic enantiomer

Fig. 11   Experimental design of a pyrethroid binding assay. (From Soderlund, 1985.)

$[^{14}C]$-*cis*-permethrin in the housefly *Musca domestica* both *in vivo*, and using isolated membrane preparations. The putative receptors showed some of the expected properties of target sites for these insecticides. A negative temperature coefficient of binding was detected, together with a degree of sensitivity to calcium. The receptor saturated at 45–90 nM (in the case of the DDT binding site), and yielded an apparent K_D of 12·2 nM. The maximum number of binding sites was 17 pmol DDT mg^{-1} membrane protein (0·34 pmol/ housefly head). Competition studies indicated that DDT, *cis*-permethrin and cypermethrin were binding to the same receptor, though not at precisely the same site. Based on these findings it was suggested that the pyrethroid membrane target was calcium-sensitive with a possible role in ion conductance.

However the highly lipophilic nature of the radioligand presents serious difficulties in applying conventional binding protocols. This was recognized by Soderlund and collaborators (Soderlund *et al.*, 1983; Soderlund, 1985) who employed an experimental approach based on that devised for stereo-specific binding to opiate receptors in an attempt to circumvent this problem (Fig. 11). By this means saturable, specific binding of $[^3H]$-NRDC 157 was measured in the presence of unsaturable and saturable non-specific binding (Fig. 12). As illustrated, the method required measurement of binding of radioligand alone (A), binding in the presence of 100-fold excess of un-labelled non-toxic enantiomers (B) and unlabelled toxic enantiomer (C).

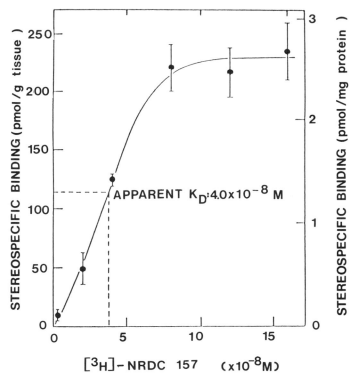

Fig. 12 Saturation curve for the binding of [³H]-NRDC 157 to mouse brain membranes. (From Soderlund *et al.*, 1983.)

Thus A and C yield total binding and unsaturable binding directly, whereas nonspecific saturable binding is measured as the difference in binding under conditions A and B. Stereospecific binding is measured as the difference in binding between conditions B and C.

Despite the considerable experimental difficulties, a minor component of the total binding appeared to be both saturable and stereospecific (Table 2). Stereospecific binding was present in both microsomal and synaptosomal fractions derived from mouse brain but was absent from the nuclear fraction prepared from brain and a microsomal fraction obtained from mouse liver (Soderlund *et al.*, 1983). Half-saturation at 40 nM [³H]-NRDC 157 (1R, *cis*) provided the only estimate for the K_D (Fig. 12). A more precise determination was precluded by the high levels of non-specific binding. Though clearly a marginal experiment because of the extreme technical difficulties acknowledged by the authors, the results provide indications that a high-affinity, stereospecific receptor-like pyrethroid binding site may be present in CNS membrane preparations.

TABLE 2   Binding of [³H]-NRDC 157 to mouse
brain membranes[a]

Type of binding	Amount (pmol mg^{-1} protein)
Stereospecific, saturable	2·54
Nonspecific, saturable	2·24
Nonspecific, saturable	95·1

[a]Table from Soderlund et al. (1983).

Vincent et al. (1980) used radiolabelled toxin II from the sea anemone
Anemonia sulcata to probe the sodium channel in rat brain synaptosomes.
The binding site characterized by this toxin was distinct from that of other
gating system toxins such as batrachotoxin, veratridine, grayanotoxin, aconitine and the pyrethroids. Jacques et al. (1980) examined the actions of a series
of pyrethroids on the sodium channel of mouse neuroblastoma cells
(N1E115), using electrophysiology and $^{22}Na^+$ flux measurements. When
tested alone, pyrethroids did not measurably stimulate $^{22}Na^+$ entry through
the Na^+ channel. However, they did stimulate $^{22}Na^+$ entry when used in
conjunction with other toxins specific for the gating mechanism of the channel. These included batrachotoxin, veratridine, dihydrograyanotoxin II and
polypeptide toxins such as anemone toxin II and scorpion toxins. This stimulating effect was fully inhibited by the sodium channel blocker tetrodotoxin
(TTX). Half-maximal saturation of the pyrethroid receptor was seen in the
micromolar range for the most active pyrethroids. The synergism between
the effects of pyrethroids on $^{22}Na^+$ flux and the actions of other toxins
showed the insecticides to be acting at separate sites on the sodium channel.
In the same study pyrethroids were found to be active on the silent sodium
channels of C9 (rat brain tumour cells) and some were able to bind without
activating $^{22}Na^+$ movement (acting as antagonists to the active pyrethroids).
   Ghiasuddin and Soderlund (1984, 1985) examined the actions of sodium
channel toxins and pyrethroids on $^{22}Na^+$ uptake into mouse brain synaptosomes. Preincubation of membranes with deltamethrin under sodium-free
conditions appeared to have no effect on the subsequent $^{22}Na^+$ uptake.
However, preincubation with deltamethrin in the presence of veratridine
enhanced the veratridine-dependent $^{22}Na^+$ uptake resulting from
veratridine-activation of sodium channels, an effect that was blocked by
TTX. These results demonstrated a specific pyrethroid action on the function
of voltage-dependent sodium channels in mouse brain synaptosomes (see
also Soderlund, 1985). The requirement for veratridine activation is consistent with electrophysiological results, which show that the rate of closing of

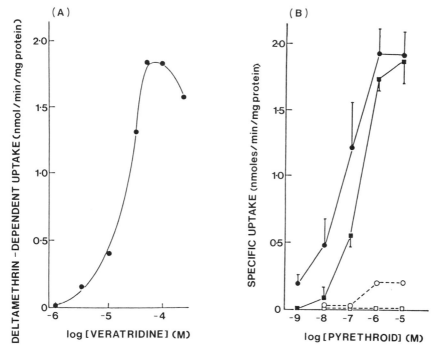

Fig. 13   (A) Effect of veratridine concentration on initial rates of deltamethrin-dependent specific $^{22}Na^+$ uptake (1.0×10^{-5} M deltamethrin). Points are means of values obtained in two replicates using different synaptosomal preparations with three determinations under each experimental condition in each replicate. (B) Effect of pyrethroid concentration on initial rates of pyrethroid-dependent specific $^{22}Na^+$ uptake in the presence of 5.0×10^{-5} M veratridine: (●) deltamethrin; (■) NRDC 157; (○) enantiomer of deltamethrin; (□) enantiomer of NRDC 157. Data points are means of values obtained in four replicates (deltamethrin and NRDC 157; bars show standard errors) or two replicates (1S enantiomers) using different synaptosomal preparations with three or four determinations under each experimental condition in each replicate. (From Ghiasuddin and Soderlund, 1985.)

activated channels is slowed by pyrethroids (see Sections 2.1.2 and 2.1.3 above).

Of the four pyrethroids examined by Ghiassudin and Soderlund (1985) deltamethrin proved the most potent (Fig. 13) in enhancing veratridine-dependent activation of $^{22}Na^+$ uptake. NRDC 157 was approximately nine-fold less potent. The insecticidally-inactive isomers of these pyrethroids had little or no effect in this flux assay. Such results confirmed the earlier work by Lazdunski and collaborators (Jacques *et al.*, 1980) on neuroblastoma cells and showed that the effects of pyrethroids on adult mammalian brain tissue are both potent and stereospecific for toxic isomers.

Similar sodium flux measurements are urgently needed on insect tissues to parallel the growing body of information on pyrethroid–sodium channel interactions obtained using electrophysiological techniques.

2.2 POTASSIUM CHANNELS

Several distinct types of potassium channels have been characterized in vertebrate and invertebrate neurones. These include delayed (outward) rectifiers, inward rectifiers, outward transients and calcium activated channels (Hille, 1984). Experiments on insect neurones have characterized a delayed (outward) rectifier in axonal membranes of giant interneurones (Pelhate and Sattelle, 1982) and expanded presynaptic regions of cholinergic sensory neurones (Blagburn and Sattelle, 1987a). Calcium-activated potassium channels have been characterized in identified neuronal cell bodies (Thomas, 1984) presynaptic regions of identified cholinergic sensory neurones (Blagburn and Sattelle, 1987b) and dissociated insect neurones (Yamamoto et al., 1987).

Minor effects of pyrethroids have been reported on potassium channels at relatively high concentrations. For example, when perfused internally in the squid axon, allethrin reduced the amplitude of the outward potassium current without affecting its kinetics (Narahashi and Anderson, 1967; Wang et al., 1972). However, tetramethrin (3.0×10^{-4} M) and fenvalerate (1.0×10^{-5} M) were without effect when tested on the potassium current of crayfish axons (Lund and Narahashi, 1979). Similarly, no clear effects of 8 pyrethroids could be detected on the potassium current of cockroach giant axons (Laufer et al., 1984). Thus the reported effects on potassium channels are unlikely to play a major role in the insecticidal actions of pyrethroids though for some species the effect of closing potassium channels at the same time as prolonging the duration of sodium channel openings may combine to increase the likelihood of repetitive firing in axons. Clearly, as a result of actions, on other cation channels (calcium and sodium), intracellular ion changes could lead to secondary effects on, for example, calcium-activated potassium channels.

2.3 CALCIUM CHANNELS

Calcium channels are able to generate transmembrane electrical signals, intracellular chemical signals or a combination of both. Recently three distinct types of calcium channel have been recognized in vertebrate tissues. The T, N and L calcium channels are recognized by steady-state inactivation, activation threshold, selectivity and pharmacology (Nowycky et al., 1985; McCleskey et al., 1986). Calcium channels have been found in insect neuronal cell bodies (Pitman, 1979; Goodman and Spitzer, 1981; Thomas, 1984) and presynaptic endings of cholinergic sensory neurones (Blagburn and Sattelle, 1987b).

No systematic study of pyrethroid actions of calcium channels has been performed, and the few reports concerning this issue are conflicting. Orchard and Osborne (1979) are of the opinion that permethrin induces repetitive firing in the neurohaemal tissue of *Carausius* (see Fig. 3) by promoting calcium channel function, because calcium appears to be the predominant charge carrier in that tissue (Orchard and Osborne, 1977). However, it has not been established that axonal action potentials are solely mediated by calcium channels, though this is the case for the cell body membrane of the same neurones (Orchard, 1976). Instead, Orchard and Osborne (1977) have shown that the amplitude of action potentials recorded extracellularly from the neurohaemal axons is reduced when the external sodium concentration is decreased, suggesting the existence of a sodium component. It should be pointed out that modification of a small fraction of sodium channels is sufficient to elevate the depolarizing after-potential to the threshold level for repetitive discharges or symptoms of poisoning at the whole animal level. In this context it should be recalled that Lund and Narahashi (1982) have calculated that less than 0·1% of the total number of sodium channels existing in the nerve is enough to produce such an effect. Therefore, the repetitive discharges observed in the permethrin-treated neurohaemal axons could not be taken as an indication of pyrethroid action on calcium channels.

A suitable preparation to examine the effects of pyrethroids on insect calcium channels is the insect skeletal muscle, as the inward current associated with excitation of muscle fibres is exclusively mediated by calcium channels (Yamamoto and Fukami, 1977; Yamamoto *et al.*, 1981; Ashcroft and Stanfield, 1982; Salkoff and Wyman, 1983). The insect muscle fibres lack sodium channels, though sodium ions can penetrate the membranes *via* the calcium channels (Yamamoto, 1987; Yamamoto and Washio, 1979). Clements and May (1977) reported that electrically excited responses as well as graded potentials superimposing on the neurally evoked excitatory postsynaptic potentials (epsps) of the locust muscle were eliminated by a Type I pyrethroid. The simplest explanation for this is that the calcium channels in the muscle membrane were blocked by the pyrethroid. On the other hand, Salgado *et al.* (1983a, b) have demonstrated that non-synaptic calcium electrogenesis in the housefly prepupae is not impared by deltamethrin. Clearly, further study is needed to solve this discrepancy.

It is of interest to recall that some hydrophobic sodium channel openers such as veratridine, grayanotoxin, and batrachotoxin were found to be potent calcium channel blockers in mouse neuroblastoma cells (Romey and Lazdunski, 1982). Whether or not this is the case for pyrethroids remains to be established. Within the last three years, considerable evidence has accumulated that three (or perhaps more) distinct types of calcium channels exist (Nowycky *et al.*, 1985). Because pharmacological properties of these subtypes are quite distinct (Miller, 1987), it will be very important to clarify which population of calcium channels is under investigation when examining pyrethroid actions. The presynaptic calcium channels in cholinergic sensory neurones of the cockroach *Periplaneta americana* (Blagburn and Sattelle, 1987b) offer another suitable preparation for the study of pyrethroid–calcium channel interactions. Preliminary evidence (Blagburn and Sattelle, 1987b) indicates that these may be N-type dihydropyridine-insensitive calcium channels.

Very recently, a voltage-independent divalent cation channel was discovered in Balb/c 3T3 cells that were activated by epidermal growth factor (EGF), and tetramethrin was found to block this channel (I. Kojima, personal communication).

3 Pyrethroid–receptor interactions

3.1 ACETYLCHOLINE RECEPTORS

Acetylcholine (ACh) receptors are regulatory membrane proteins which bind ACh released from presynaptic neurones at cholinergic chemical synapses (Heidmann and Changeux, 1978). Synapses at which ACh is the neurotransmitter enable chemical communication (a) between neurones, (b) between neurones and gland cells, and (c) between neurones and muscle cells. ACh receptors may be located postsynaptically, mediating directly the interaction of pre- and post-synaptic cells. Presynaptically-located ACh receptors can modulate the pathway by influencing ACh release. Finally, extrasynaptically-located ACh receptors have been detected on vertebrate and invertebrate neurones, though their functions are less clear.

Distinct differences in their responses to drugs have led to the classification of ACh receptors into separate nicotinic and muscarinic categories: nicotinic ACh receptors of muscle are activated by ACh and blocked by d-tubocurarine and α-bungarotoxin, whereas those in ganglia and CNS show slight differences in pharmacology—hexamethonium and κ-bungarotoxin often being the most effective antagonists; muscarinic receptors are blocked by atropine and quinuclidinyl benzilate. The well-characterized vertebrate peripheral nicotinic ACh receptor is a multimeric protein with 4 subunits in the molar ratio $\alpha_2\beta\gamma\delta$ (Conti-Tronconi and Raftery, 1982; Takai *et al.*, 1985) and the complete amino acid sequence of all four *Torpedo* subunits has been deduced from their complementary DNAs (Noda *et al.*, 1982, 1983a, b; Claudio *et al.*, 1983).

More recently, cDNA encoding the muscarinic acetylcholine receptor was also cloned, and the primary structure of the receptor protein was determined (Kubo *et al.*, 1986). Unlike the nicotinic receptor, the muscarinic receptor was a single polypeptide composed of 460 amino-acid residues, having approximately 30% homology with the β-adreno receptor and rhodopsin.

Both nicotinic and muscarinic ACh receptors are present in insect CNS, in a ratio of approximately 10:1 (Sattelle, 1985, 1986). Insect CNS is a rich source of neuronal nicotinic receptors for which a postsynaptic role has been identified (Sattelle *et al.*, 1983). Presynaptic (Blagburn and Sattelle, 1987c) and extrasynaptic (David and Sattelle, 1984) insect nicotinic receptors have been located on identified neurones. A functional role for presynaptic muscarinic receptors has also been established in insect CNS (Breer and Knipper, 1984). ACh receptors are shown to be sites of action, and in some cases primary molecular targets, for a variety of insecticides such as nicotine, cartap (= nereistoxin), aminocarb and aldicarb (Eldefrawi, 1976; Edefrawi *et al.*, 1982; Sattelle and Callec, 1977; Sattelle *et al.*, 1985). Initial studies indicate some biochemical differences between the α-bungarotoxin binding component of insect CNS and vertebrate muscle, though the insect receptor is clearly multimeric (Breer *et al.*, 1985; Sattelle and Breer, 1986). In this section recent work on the actions of pyrethroids on nicotinic ACh receptors is considered.

3.1.1 *Electrophysiology*

Electrophysiological studies on the effects of pyrethroids on skeletal muscle neuromuscular transmission in vertebrates revealed distinct postsynaptic facilitation, leading to the suggestion that pyrethroids have no inhibitory postsynaptic action (Evans, 1976; Wouters *et al.*, 1977). More recently, Sherby *et al.* (1986) showed that the ACh sensitivity of denervated soleus muscle fibres was greatly depressed when a low-sodium, high-calcium buffer was employed, but fluvalinate ($1 \cdot 0 \times 10^{-5}$ M) significantly increased ACh sensitivity. On the basis of these findings it was suggested that pyrethroids bind to the nicotinic ACh receptor and interfere with receptor desensitization.

A number of laboratories have now examined the actions of pyrethroids on invertebrate nicotinic ACh receptors. Leake (1982) working on Retzius neurones of the leeches *Hirudo medicinalis* and *Haemopsis sanguisuga* showed that these cells were excited by carbamylcholine, a response that was blocked by $5 \cdot 0 \times 10^{-4}$ M benzoquinonium. A number of pyrethroids were shown to induce depolarizations on these cells, and pretreatment of the preparation with benzoquinonium failed to reduce the stimulating actions of pyrethroids. It was concluded that the pyrethroids tested including

S-bioallethrin, were not acting on the ACh receptors. Whilst it seems clear that an action at ACh receptors does not contribute to the depolarizing actions of pyrethroids on these cells, the possibility that pyrethroids exert, for example, a blocking action at the leech ACh receptor cannot be ruled out by these studies.

A synaptic blocking action of deltamethrin has been demonstrated in insect CNS at the cercal afferent, giant interneurone synapse of the cockroach *Periplaneta americana* (Hue, 1987). Using the oil-gap, single fibre recording method, deltamethrin at a concentration of $1\cdot0 \times 10^{-5}$ M produced a parallel block of both the epsp and the response of giant interneurone 2 to ionophoretically-applied ACh. Though there was some indication of a presynaptic action, the basis of which requires further elucidation, the evidence for a postsynaptic blocking action, probably at the level of the postsynaptic nicotinic receptor, is unequivocal. In a voltage-clamp study (Sattelle, 1988) a direct demonstration has been provided of a blocking action of deltamethrin at an insect nicotinic ACh receptor. The cell body membrane of an identified insect motor neurone (the cockroach fast coxal depressor) was used to record ACh-induced currents in the presence and absence of $1\cdot0 \times 10^{-6}$ M deltamethrin. A clear suppression of the ACh-induced current was demonstrated. Since this cell does not receive synaptic inputs on to the cell body membrane, the actions on the cell body nicotinic receptors are not complicated by possible presynaptic actions of the pyrethroid.

3.1.2   *Biochemical studies*

Biochemical studies have shown that pyrethroids influence the binding of radiolabelled ligands to the nicotinic ACh receptor of the electric organ of the electric ray *Torpedo*. The pharmacology of this receptor is similar to that of mammalian skeletal muscles. Pyrethroids do not inhibit the binding of [3H]-ACh to these nicotinic receptor sites but they were found to inhibit the binding of radiolabelled perhydrohistrionicotoxin ([3H]-HTX) to the allosteric channel sites (Abassay *et al.*, 1982, 1983a, b). Pyrethroids were more active on [3H]HTX binding sites at low temperatures, and also inhibited receptor-activated $^{45}Ca^{2+}$ uptake. Also, Type I pyrethroids were more potent and faster acting than Type II pyrethroids. These studies indicated that pyrethroids affected nicotinic receptor function, though possibly at higher concentrations than their effects on the axonal Na^+ channel. It was clear from the work of Abassy *et al.* (1983b) that the alcohol moiety of pyrethroids was particularly important for their actions on nicotinic ACh receptors. Esters of cyclopentenolone (allethrin and bioallethrin) of Esbiol (RU 19 177 and RU 40 246), of 5-benzyl-3-furylmethyl alcohol (kadethrin and resmethrin), and of tetrahydrophthalimidomethyl alcohol (tetramethrin) were

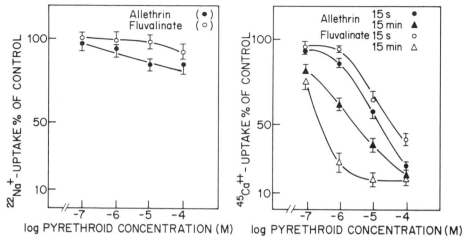

Fig. 14 Effects of allethrin (●) and fluvalinate (○) on Naja α-neurotoxin-sensitive, carbamylcholine-stimulated tracer ion uptake. (A) $^{22}Na^+$ uptake. *Torpedo* microsacs were preincubated (at 2°C) with the pyrethroid for 15 min before exposure to $^{22}Na^+$ for 60 s. (B) $^{45}Ca^{2+}$ uptake. *Torpedo* membranes were exposed to pyrethroid either simultaneously with $^{45}Ca^{2+}$ for 15 s (○,●) or preincubated (at 2°C) with the pyrethroid for 15 min (▲,△) before exposure to $^{45}Ca^{2+}$ for 15 s. Symbols and vertical bars represent means ± the standard deviation of three experiments. (From Sherby *et al.*, 1986.)

more potent in inhibiting carbamylcholine stimulated [³H]-HTX binding than the esters of α-cyano-3 phenoxybenzyl alcohol. Changes in the acidic moiety had little effect. An apparent anomaly in this earlier work recognized by the authors was that pyrethroids were more potent (10–100 fold) in inhibiting $^{45}Ca^{2+}$ flux than they were in inhibiting [³H]-HTX binding.

In an attempt to resolve this Sherby *et al.* (1986) examined the actions of two pyrethroids using both ligand binding and physiological methods. Allethrin and fluvalinate were inactive on the equilibrium binding of [^{125}I]-α-bungarotoxin to the *Torpedo* nicotinic receptor, but both decreased its initial rate of binding in the presence of the cholinergic agonist carbamylcholine. Binding of [³H]-HTX to the channel site of the receptor was inhibited more by allethrin than by fluvalinate, but unlike perhydrohistrionicotoxin neither pyrethroid had much effect on receptor-activated $^{22}Na^+$ influx into *Torpedo* microsacs (Fig. 14A), nor did fluvalinate reduce ACh sensitivity in denervated rat soleus muscle. Thus it was suggested that the pyrethroids were binding to a site distinct from either the agonist site or the perhydrohistrionicotoxin binding site. Both fluvalinate and allethrin were very active in inhibiting receptor-activated $^{45}Ca^{2+}$ uptake (Fig. 14B) and their effectiveness increased with time. The RS fluvalinate isomer was more potent than the

RR isomer in inhibiting $^{45}Ca^{2+}$ uptake in line with the observed relative toxicity to whole insects. Fluvalinate and allethrin ($1\cdot0 \times 10^{-6}$ M) increased the affinity of nicotinic ACh receptors for carbamylcholine at low concentrations and decreased it at high concentrations, suggesting that binding of pyrethroids affects agonist-induced conformational changes.

Though the precise site of action remains to be determined, and to date biochemical studies on pyrethroid-receptor interactions in insects are lacking, there is now sufficient evidence that pyrethroids are active at nicotinic ACh receptors of vertebrates and insects. Block of physiological responses to ACh have been demonstrated and ligand binding and flux measurements indicate modification of agonist-induced conformational changes of the receptor. Since most actions are detected at approximately micromolar concentrations and actions at this site do not account for many of the observed effects of pyrethroids, it is likely that the nicotinic ACh receptor represents an additional rather than a primary site of action of these insecticides.

3.2 GABA RECEPTORS

There is abundant evidence that GABA is a major inhibitory neurotransmitter in vertebrates and invertebrates (Kuffler and Edwards, 1958; Usherwood and Grundfest, 1965; Otsuka et al., 1966; Roberts et al., 1976). Much of the recent interest in insect nervous system, GABA receptors has stemmed from the demonstration that insecticidally-active molecules including α-cyano-containing pyrethroids (Lawrence and Casida, 1983), avermectins (Campbell et al., 1983) and cyclodienes (Lawrence and Casida, 1983) appear to be effective at this membrane site in the mammalian central nervous system (CNS). There is growing evidence for pharmacological differences between insect and vertebrate CNS receptors. The CNS of vertebrates contains two distinct subclasses of GABA receptors. The GABA$_A$ receptor is linked directly to a chloride ion channel (Enna and Gallagher, 1983; Gallagher and Shinnick-Gallagher, 1983). Bicuculline is a specific antagonist of vertebrate GABA$_A$ receptors where it binds to the same recognition site as GABA (Olsen, 1981). The GABA$_B$ receptor is less well understood, but baclofen is clearly a specific agonist (Simmonds, 1983), and there is evidence for a quite different mechanism of activation. Electrophysiological evidence points to a direct modification of calcium channel conductance (Dunlap and Fischbach, 1981) and/or potassium conductance (Newberry and Nicoll, 1984a, b). There is some evidence that GABA$_B$ receptors may act indirectly via adenylate cyclase (Dolphin, 1984).

Specific saturable binding of [^{3}H]-GABA and [^{3}H]-muscimol have been demonstrated for several insect species. These specific binding sites in the cockroach (Lummis and Sattelle, 1985a, 1986), the locust (Breer and Heilgenberg, 1985; Lunt *et al.*, 1985) and the honeybee (Eldefrawi *et al.*, 1985) differ from the GABA binding sites characterized in vertebrates as they are inhibited neither by bicuculline nor by baclofen. The location of specific [³H]-GABA binding in the neuropile of insect ganglia is consistent with a synaptic role for this putative receptor (Lummis and Sattelle, 1985a). The recent finding that a GABA-activated Cl⁻ uptake into microsacs prepared from *Periplaneta* CNS is insensitive to bicuculline (Wafford *et al.*, 1987) and that GABA responses on an identified neurone in the same tissue are not blocked by bicuculline (Wafford *et al.*, 1987; Sattelle *et al.*, 1988) confirms a functional role in insects for a bicuculline-insensitive GABA receptor that activates a Cl⁻ channel. Other classes of GABA receptor may prove to be present in insect CNS, but studies to date of pyrethroid–GABA receptor interactions appear to relate to this particular receptor.

The mammalian central nervous system GABA$_A$ receptor/chloride channel complex contains binding sites for [³H]-dihydropicrotoxin (Leeb-Lundberg and Olsen, 1980), benzodiazepines such as [³H]-flunitrazepam (Braestrup and Squires, 1977) and bicyclophosphates such as [³⁵S]-t-butylbicyclophosphorothionate (Squires *et al.*, 1983; Supavilai and Karobath, 1984). Similar sites are present in insect CNS, though evidence from several laboratories indicates that both the benzodiazepine site (Lummis and Sattelle, 1986; Robinson *et al.*, 1986) and the TBPS site (Cohen and Casida, 1986) differ in their pharmacology from the corresponding sites in vertebrate CNS.

3.2.1   *Electrophysiology*

Following up the intriguing toxicological observations that a benzodiazepine (diazepam) exerted a protective effect on the actions of deltamethrin (though not permethrin) in the mouse and the cockroach (Gammon *et al.*, 1981), Gammon and Casida (1983) showed that deltamethrin and 3 insecticidal cyano analogues increased the input resistance of crayfish (*Procambarus clarkii*) claw opener muscle fibres bathed in GABA. Non-toxic stereoisomers and compounds that result in Type I symptoms of poisoning were inactive. Benzodiazepines reduced the potency of both deltamethrin and the GABA antagonist picrotoxinin. It was therefore suggested that the cyanophenoxybenzyl pyrethroids were active on the crustacean GABA receptor/chloride channel complex. Similar conclusions were reached in the extension of this study (Gammon, 1985; Gammon and Sander, 1985) in which the effects of pyrethroids on arthropod GABA receptors and sodium channels from different preparations were compared.

More recently, using the same species of crayfish, Chalmers *et al.* (1987)

compared the actions of pyrethroids on the stretch receptor neurone. This preparation enables a direct comparison on the same tissue of the capacity of the applied compound to modify sodium channels and GABA receptors that activate Cl^- channels. The lowest concentration of deltamethrin to exert an effect on sodium channels was 1.0×10^{-12} M, whereas the response of the preparation to GABA appeared to be unaffected by deltamethrin concentrations up to 1.0×10^{-7} M. Although deltamethrin at 1.0×10^{-6} M slightly reduced the response to GABA, the authors' conclusion was that the majority of the actions of cyano pyrethroids on the crayfish preparation could be explained in terms of their actions on the sodium channel.

Until recently physiological data on the target group of invertebrate organisms has been lacking. Sattelle (1988) working on an identified insect motor neurone (the fast coxal depressor motorneurone of the cockroach *Periplaneta americana*) has shown that GABA-induced currents on the cell body membrane of this neurone are unaffected by deltamethrin concentrations in the range 1.0×10^{-8} M to 5.0×10^{-7} M (Fig. 15), though a blocking action appears at concentrations above 1.0×10^{-6} M. Thus in insects the physiological evidence available to date does not support the role of an insect GABA-activated chloride channel as the primary site of action of this pyrethroid. Electrophysiological studies to date on vertebrates have not examined directly the actions of pyrethroids on the GABA response of a single neurone. Nevertheless, studies on spinal ganglia have been performed. Smith (1980) concluded that cismethrin enhancement of dorsal root potentials did not result from block of GABA-mediated inhibitory synaptic transmission. Staatz-Benson and Hosko (1986) showed that a cyano-substituted (deltamethrin) and a non-cyano substituted (*cis*-permethrin) pyrethroid both enhanced spontaneous firing rates of rat ventral horn interneurones. The same authors showed that, in the cat, diazepam was equally effective in antagonizing facilitation produced by either deltamethrin or *cis*-permethrin.

3.2.2   *Biochemical studies*

Radiolabelled ligand binding experiments have been used to characterize in detail putative GABA receptors of vertebrates (Fischer and Olsen, 1986) and to examine the actions of pyrethroid insecticides on putative GABA receptors of vertebrates and insects. In addition to [³H]-GABA and [³H]-muscimol which probe the recognition sites of GABA receptors, other radioligands have been employed. [³H]-bicuculline binds to the receptor site in vertebrates, though the lack of effect of bicuculline in most physiological studies on insects may explain why, to date, it has not been used on insect preparations. [³H]-Flunitrazepam and [³H]-Ro 5-4864 bind respectively to the central and peripheral benzodiazepine receptor sites of vertebrates and

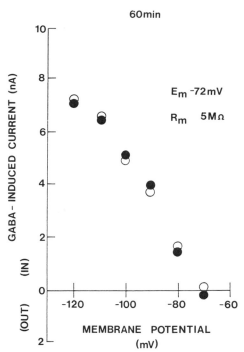

Fig. 15 Absence of a blocking effect of deltamethrin (5.0×10^{-7} M) on GABA-induced current recorded at various membrane potentials from the cell body membrane of the cockroach fast coxal depressor motor neurone. (●), control, (○), deltamethrin 5.0×10^{-7} M. The resting potential of the cell (E_m) and its input resistance (R_m) prior to the voltage-clamp experiment are indicated. (From Sattelle, 1988.)

[^{35}S]-TBPS ([^{35}S]-t-butylbicyclophosphorothionate) and [^{3}H]-dihydropicro-toxinin appear to act at other sites, possibly related to the GABA receptor ionic channel. Coupling between such sites and the GABA receptor is demonstrated by the finding that GABA influences the binding of these ligands to their sites of action. Whereas, in the case of Ro 5-4864, coupling to a GABA receptor has not been demonstrated in vertebrates, this ligand is the most effective inhibitor of [^{3}H]-flunitrazepam binding to insect nervous tissue.

The earliest *in vitro* study showing that cyanophenoxybenzyl pyrethroids interact with a site on the GABA-activated Cl$^-$ channel was the demonstration by Leeb-Lundberg and Olsen (1980) that deltamethrin (IC$_{50}$, 100 nM) displaced [^{3}H]-dihydropicrotoxinin binding to rat brain synaptosomes. The insecticidally-inactive αR analogue did not affect this binding site indicating stereospecificity of the action. However, this particular radioligand binds with a very low signal-to-noise ratio, limiting detailed structure-activity studies.

The discovery by Squires *et al.* (1983) of the selective radioligand [^{35}S]-TBPS, which was effective at the picrotoxin site on the mammalian GABA-receptor, enabled further progress. In a detailed investigation by Lawrence and Casida (1983), 37 pyrethroids were tested on [^{35}S]-TBPS binding to rat brain synaptic membranes. The binding assays established structure-activity relations *in vitro* that agreed well with structure-toxicity relations in mice. The most effective pyrethroids were 1R, *cis*, αS cypermethrin, deltamethrin and 2S, αS fenvalerate (see Fig. 15a) and the interaction appeared to be allosteric or non-competitive (Lawrence and Casida, 1983; Lawrence *et al.*, 1985).

Lawrence and collaborators (1985) showed that the binding of [^{3}H]-Ro 5-4864 was also inhibited by cyano-containing pyrethroids (see Fig. 15b). Of interest was the demonstration that cypermethrin was 60 × more potent on this site compared to the TBPS site. Deltamethrin and 1R, *cis*, αS cypermethrin were the most potent of the pyrethroids examined. Two pyrethroids with typical Type I activity (permethrin and allethrin) were relatively weak inhibitors of [^{3}H]-Ro 5-4864 binding.

Thus several lines of experimental evidence have accumulated, largely from mammalian sources, that the GABA receptor/chloride channel may be a key site of pyrethroid action. The onset of Type II, but not Type I symptoms are delayed by diazepam treatment of cockroaches and mice and the signs of toxicity resemble those observed for other known GABA antagonists. In addition, the specific binding of radioligands known to interact with the GABA receptor/chloride channel is inhibited in a stereospecific fashion by pyrethroids with Type II action. In the case of [^{35}S]-TBPS and [^{3}H]-Ro 5-4864, the ability of pyrethroids to inhibit binding is closely related to their mammalian toxicities.

Lummis and colleagues (1987a) examined the actions of 1.0×10^{-4} M deltamethrin on the binding of [^{3}H]-muscimol, [^{3}H]-diazepam and [^{35}S]-TBPS to rat brain membranes. This pyrethroid reduced by 20% the specific binding of [^{35}S]-TBPS, and suppressed by 30% the specific binding of [^{3}H]-diazepam. In contrast, the specific binding of [^{3}H]-muscimol was increased by 60%. Modification of all these three binding sites points to an interaction with the receptor complex, albeit at high concentrations. Examination of K_D and B_{max} data revealed that none of these actions of deltamethrin was competitive. M. E. and A. T. Eldefrawi have characterized GABA receptors in rat brain by radioligand binding, and using GABA-activated ^{36}Cl$^-$ influx into rat brain microsacs (Abalis *et al.*, 1983a, 1986; Eldefrawi *et al.*, 1985). At micromolar concentrations α-cyano-containing pyrethroids produced a partial block of GABA activated ^{36}Cl$^-$ uptake into rat brain microsacs (Table

TABLE 3   Effects of pyrethroids on GABA-induced $^{36}Cl^-$ uptake into rat brain microsacs

Insecticide	$^{36}Cl^-$ uptake[a] (nmol/mg protein)	% of control
Allethrin	18.0 ± 0.9	83
Fluvalinate	12.0 ± 0.3	56
1R,cis oS-Cypermethrin	5.9 ± 0.7	28
1R,trans oS-Cypermethrin	13.7 ± 1.1	64

[a] Each value is the mean of three experiments ± S.D. Data from Abalis *et al.* (1986).

3). The same laboratory has performed a detailed characterization of the housefly GABA/benzodiazepine receptors (Abalis *et al.*, 1983b). Further studies of pyrethroid actions on this receptor will be of considerable interest.

The usefulness of $[^{35}S]$-TBPS as a putative receptor ligand in many verte-brate species (Wong *et al.*, 1984; Cole *et al.*, 1984) has led to attempts to solubilize this binding site from rat brain tissue (Seifert and Casida, 1985). Using the zwitterionic detergent CHAPS (3-((3-cholamidopropyl) dimethylammonio)-1-propanesulfonate), the binding of both GABA and cypermethrin to the $[^{35}S]$-TBPS site are modified (Seifert and Casida, 1985a, b). Recently, specific binding of $[^{35}S]$-TBPS to a housefly thorax + abdomen preparation has been characterized (Cohen and Casida, 1986). It is of interest that these authors find considerable differences between this site in housefly membranes and its counterpart in mammalian brain with respect to many GABAergic agents, especially in view of other reported differences in the pharmacology of insect and mammalian GABA receptors linked to Cl channels (see Section 3.2.1).

As new receptor assays emerge it is appropriate to re-examine the pharma-cology of a particular site especially if the new method can be more closely related to the normal functional role of the receptor under investigation. The development of GABA-activated $^{36}Cl^-$ uptake assays for mammalian (Harris and Allan, 1985) and insect (Wafford *et al.*, 1987) CNS tissues repre-sents just such an advance. Using this assay, a wide range of drugs and toxic agents including pyrethroid insecticides have been shown to exert some action on the GABA-activated uptake of $^{36}Cl^-$ into membrane vesicles pre-pared from mammalian brain. By this means Bloomquist and Soderlund (1985) and Bloomquist *et al.* (1986) showed that deltamethrin (2.5×10^{-5} M) resulted in about 50% inhibition of GABA-dependent Cl$^-$ uptake, but the extent of block was not increased at higher concentrations of the pyrethroid. The low potency and incomplete stereospecificity of deltamethrin inhibition led these authors to suggest that the GABA receptor/chloride channel is not involved in the primary neurotoxic actions of α-cyano-substituted pyrethroids. Recently, using membrane vesicles prepared from the cockroach (*Periplaneta americana*) nervous system, deltamethrin (1.0×10^{-5} M) was shown to block GABA-activated $^{36}Cl^-$ uptake (Lummis *et al.*, 1987b), though the effects of lower concentrations, known not to affect physiological responses to GABA in the same species (Sattelle, 1987b), have not yet been examined in the $^{36}Cl^-$ flux assay.

Thus the view that all the effects of α-cyano-containing (Type II) pyrethroids can be explained by a separate site of action (the GABA receptor/chloride channel) from that of the non-α-cyano containing (Type I) pyrethroids (the sodium channel) may require qualification. It does not seem to hold for insects, and contradictory data both for and against this view has been obtained from crustacean preparations. Though pyrethroids containing an α-cyano group do undoubtedly act at several sites on mammalian GABA receptors, their capacity to modify GABA-activated Cl^- uptake is only seen at high concentrations and is unlikely to represent a primary site of pyrethroid action. Should TBPS prove to be an anionic channel probe rather than a selective probe of only GABA-activated Cl^- channels, some of the discrepancies currently evident in the literature could perhaps be explained. However, it may be premature to discard this site completely, especially in the context of mammalian toxicology. The extensive work of Casida, Gammon and colleagues has (a) yielded an important new radio ligand, (b) advanced our understanding of the GABA receptor chloride channel, and (c) has succeeded in alerting other workers to the possibilities of multiple sites of action of pyrethroid insecticides.

4   Pyrethroid–ATPase interactions

Clark and Matsumura (1982) showed that allethrin inhibited Ca^{2+}-ATPase activity in squid nerves. Deltamethrin and cypermethrin were also effective. This blocking activity was not correlated with the presence of an α-cyano group, since permethrin (lacking an α-cyano group) was effective, whereas fenvalerate an α-cyano containing pyrethroid (like deltamethrin) was not very active (see Fig. 16).

Jones and Lee (1986) reported the effects of pyrethroids on the activity of the $Ca^{2+}-Mg^{2+}$-ATPase purified from rabbit muscle sarcoplasmic reticulum. Deltamethrin caused a small increase in the activity of the native ATPase, but a large stimulation for the ATPase reconstituted into bilayers of the short-chain phospholipid dimyristoleoyl phosphatidylcholine. The extent of stimulation was dependent on the structure of the pyrethroid. Stimulation showed a negative temperature coefficient. Pyrethroids appeared to have no

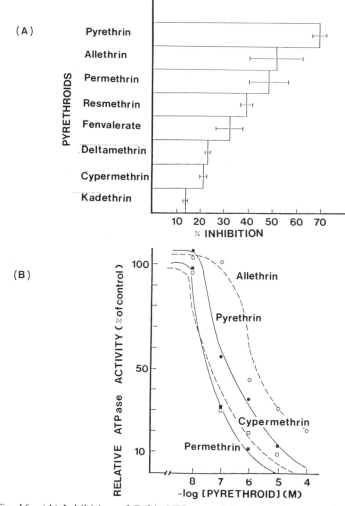

Fig. 16 (A) Inhibition of Ca^{2+}-ATPase activity by various pyrethroids. All pyrethroid solutions are adjusted to give a final assay concentration of 1.0×10^{-4} M. Percentage inhibition refers to calcium-stimulated activity only. Enzyme source is the $90,000 \times g$ fraction of the squid retinal axon preparation. ATP was added at a concentration of 5.6×10^{-8} M. Ouabain concentration is 1.0×10^{-4} M. Incubation medium was as follows: 160 mM Na^+, 160 mM K^+, 1 mM free Ca^{2+}, and 1 mM EDTA in 30 mM Tris–HCl at pH 7.1. Mean values ± S.E. are shown. (B) Effect of various pyrethroids on Ca^{2+}–Mg^{2+}-ATPase activity from squid optic lobe synaptosomes. ATPase activity is reported on a relative basis as the percentage of calcium-stimulated activity of an untreated control value (i.e., no insecticide added) taken as equal to 100% activity. Ouabain concentration was 1.0×10^{-4} M, and ATP was added at a concentration of 5.6×10^{-8} M. Pyrethroids used are pyrethrin (●), allethrin (○), permethrin (■), and cypermethrin (□). The overall specific activity of Ca^{2+}–Mg^{2+}-ATPase activity from optic lobe synaptosomes is 0.15 ± 0.02 nmol P_i/mg protein/10 min. (From Clark and Matsumura, 1982.)

effect on the fluidity of lipid bilayers. The effects were therefore explained in terms of binding of pyrethroids to the ATPase.

The same authors introduced the ester of pyrene-1-methanol with cis-2,2-dimethyl-3(2,2-dichlorovinyl) cyclopropane carboxylic acid (cis-permethrin acid) as a fluorescent analogue of the pyrethroids. Relative intensities of the vibronic bands in the fluorescence emission spectrum of the ester in lipid bilayers were consistent with the location of the pyrene group in the glycerol backbone region of the bilayer. It was suggested that pyrethroids adopted a "horseshoe" conformation in the bilayer with the ester group at the lipid-water interface. Fluorescence data were interpreted in terms of one binding site per four lipids. In the Ca^{2+}–Mg^{2+}-ATPase system prepared from muscle sacroplasmic reticulum, the ester was found to bind both to sites in the lipid and on the ATPase. There are a large number of sites for the ester on the ATPase. For the ATPase reconstituted with the short chain lipid dimyristoleoyl phosphatidylcholine, the ester caused an increase in activity comparable to that produced by the pyrethroids.

Since Ca^{2+}–Mg^{2+}-ATPase is involved in sequestering calcium, it may be that this site contributes to synaptic actions of pyrethroids, but Salgado et al. (1983a, b) provided evidence that the increase in spontaneous transmitter release by pyrethroids is due to presynaptic terminal depolarization by a TTX-sensitive Na^+ influx.

An important study in this context is that of Fujita and colleagues (see Doherty et al., 1985) who tested 25 synthetic pyrethroids on the ATP-dependent uptake of $^{45}Ca^{2+}$ by crayfish (Procambarus clarkii) homogenates. A paralleled series of experiments using 20 pyrethroids was performed by applying the same type of assay to homogenates prepared from lobster (Panulirus japonicus) axons (Fig. 17). A parabolic relation was found for inhibition of $^{45}Ca^{2+}$ uptake as a function of lipophilicity in both species. No direct correlations were demonstrated between the effectiveness of the pyrethroids in inhibiting calcium uptake and their capacity to (a) increase the frequency of spontaneous discharges in crayfish nerve cords, (b) induce repetitive firing in cockroach neurones and (c) kill cockroaches. It was concluded that although some synthetic pyrethroids are moderately potent inhibitors of calcium uptake into nerve cord and axonal preparations (i.e. I_{50} for trans-resmethrin $= 1.0 \times 10^{-6}$ M) this inhibition alone could not be related to neurophysiological changes in isolated nerve preparations nor to the toxicity of these agents in insects.

Other ATPases have been examined in less detail. In 1984, Bakry et al. reported that cypermethrin, deltamethrin, fenpropathrin, cyfluthrin, fenvalerate and permethrin stimulated mitochondrial Mg^{2+}-ATPase activity from Spodoptera littoralis at concentrations in the range 1.0×10^{-7} M to 5.0×10^{-5} M, but concluded that this enzyme was not a primary site of pyrethroid

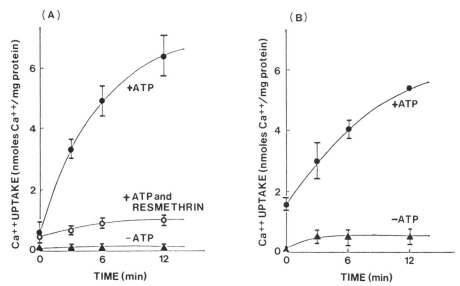

Fig. 17   Time-course of calcium uptake into crayfish central nerve cord (A) and lobster axon (B) homogenates. Data are obtained in the presence (●) and absence (▲) of 1.0×10^{-3} M tris ATP. In (A) the time-course of calcium uptake is also shown in the presence of 5.0×10^{-6} M resmethrin (○). Vertical line shows the standard deviation for three determinations. (From Doherty *et al.*, 1986.)

action. Much earlier Desaiah *et al.* (1972) showed that pyrethrins inhibited $Na^+\text{–}K^+$-ATPase from *Periplaneta americana* nerve cord and mitochondrial Mg^{2+}-ATPase at concentrations of the order of 1.0×10^{-5} M. In all such studies, relatively high concentrations are required and there is no compelling evidence to detract from the view that these are alternative rather than primary actions of pyrethroid insecticides.

5 Conclusions

5.1   THE PRIMARY SITE OF ACTION

The symptoms of pyrethroid poisoning point clearly towards an action on the nervous system. For a complete understanding of pyrethroid actions it is necessary to identify the primary molecular target and then show how these primary effects lead, possibly via a complex wave of secondary effects, to the death of the insect. There is no doubt that pyrethroids interact with sodium channels to prolong the opening of individual channels. This action appears to account for the bulk of the symptoms of pyrethroid poisoning and can explain many of the actions seen on central and peripherally located neurones. To date this is the best candidate for the primary site of action of pyrethroid insecticides.

Extending the duration of sodium channel opening will have a variety of consequences. These include prolongation of the nerve terminal depolarization, resulting in enhanced and inappropriate release of neurotransmitters and neurohormones. Changes in intracellular ions and the extraneuronal ionic environment will also follow the enhanced firing of neurones in the intact nervous system. The overall result of these secondary actions of pyrethroids will be amplification of the disruptive actions on neuronal function, but changes of this kind can all be traced back to a primary action on the sodium channel.

5.2   ROLE OF MULTIPLE TARGETS IN INSECTICIDAL ACTION

There is a growing body of evidence that pyrethroids act on membrane molecules other than sodium channels. These alternative sites of action include cation channels (calcium channels and potassium channels), GABA-activated Cl^- channels and Ca^{2+}-ATPase. There is some evidence that the GABA-activated Cl^- channels of mammalian CNS may be a site of action of the α-cyano-containing insecticides, though similar channels in the nervous system of insects are only modified at very high concentrations of pyrethroids. Actions on calcium channels and the role of Ca^{2+}-ATPase in pyrethroid actions are not firmly established, and remain to be further evaluated. All these sites appear less sensitive to pyrethroids than the sodium channel.

5.3   TOWARDS A MOLECULAR BASIS OF RESISTANCE

Genetic evidence points to a single mode of action of pyrethroids and is compatible with a single site of action on the sodium channel molecule of insect neuronal membranes. One gene, or an isolated locus, is capable of conferring site insensitivity. However, the gene product of the *kdr* gene and its sphere of influence is not yet known. Nerves of *kdr* resistant flies are clearly less sensitive to pyrethroids. This could be due to a modification of the channel protein itself, a change in its immediate lipid environment and/or an overall reduction in channel density in pyrethroid-resistant flies. Elucidating this problem presents an exciting challenge for insecticide chemists, insect physiologists, biochemists and geneticists and this important problem appears to be approachable with modern experimental techniques.

5.4   PROSPECTS FOR THE FUTURE

At this stage it is perhaps appropriate to return to the identification by

Michael Elliott in his 1985 paper (referred to in Section 1.1) of the need for an improved understanding of the nervous system molecular targets of pyrethroid insecticides, and to examine the extent to which it is being taken up and the prospects for success. It is encouraging to see that many of the references cited in this review are in fact dated after 1985 reflecting a degree of progress. The urgency for acquisition of such basic information cannot be overstated, with the spectre of pyrethroid resistance fast emerging. Recent technical advances in neuroscience that are improving dramatically our undertanding of the molecular structures of ion channels and receptors should prove applicable to the detailed characterization of the site of pyrethroid action.

For example, recent major advances in recombinant DNA and cloning methods have enabled the complete primary sequence of the sodium channel molecule of vertebrates to be elucidated. This major breakthrough yielded cDNA probes by means of which insect sodium channels were investigated (Salkoff et al., 1987). Knowledge of the primary sequence of the insect sodium channel could prove a major step forward in the study of pyrethroid-target interactions, and knowledge of the sequences for the channel in susceptible and kdr resistant houseflies would also focus the search for the molecular basis of resistance.

More detailed electrophysiology and biochemistry can be expected on susceptible and resistant strains of housefly. Now that viable, dissociated adult insect neurones, including fly neurones, can be obtained (Sattelle et al., 1986; Pinnock and Sattelle, 1987), patch-clamp studies of channels and receptors in susceptible and resistant flies are feasible. Thus nerve insensitivity to pyrethroids can probably be explored at the single channel level in the near future. In addition to exploring sodium channel kinetics, sodium channel density should be examined using a range of radioligands in normal and resistant species.

Channel reconstitution into artificial membranes, explored so successfully in the case of insect acetylcholine receptors (Hanke and Breer, 1986), could perhaps be applied to the insect sodium channel. It will then be possible to study pyrethroid-channel interactions under conditions where the lipid environment of the channel can be manipulated.

High resolution localization of radiolabelled pyrethroids following topical application, with the aim of assessing their route of access to and accumulation by different tissues, is likely to remain difficult in view of the high mobility of these insecticides in lipid environments. An antibody prepared to s-bioallethrin conjugated to haemocyanin and BSA with a view to providing a diagnostic test (Wing et al., 1978) has not to date been applied to immunocytochemical localization studies, but similar problems are likely to limit such applications.

Genetic analysis will undoubtedly continue to be used to investigate the properties and functions of *Drosophila melanogaster* ion channels (Ganetzky and Wu, 1986; Hall *et al.*, 1986). Another interesting approach to the study of pyrethroid-channel interactions and the basis of resistance might be to ask whether or not a *Drosophila* mutant such as the nap (no action potential mutant) with a reduced sodium channel density shows an altered sensitivity to pyrethroids. If channel density is a key factor in pyrethroid resistance such an effect might be predicted.

A particularly important advance is likely to be the application of electro-physiological studies on pyrethroid actions to a much wider variety of insect species than was hitherto considered possible. This is a likely outcome of the successful application of patch-clamp techniques to insect cells, and will bring to bear on species of considerable practical importance advanced physiological methods probing pyrethroid-target interactions at the single channel level.

In future, not only, will pyrethroids continue to be applied in a variety of agricultural and public health applications, but increasingly they may be combined with other compounds including biologically-engineered agents. Already a successful combination of *Bacillus thuringiensis* toxin and a sub-lethal dose of pyrethroid has been shown to be an effective control mixture for larvae of *Operophthera brumata* and *Tortrix viridana* (Svetska and Vankova, 1980). A detailed understanding of synergistic actions of pyrethroids and other compounds will provide further challenges to those seeking to explain in molecular terms how these diverse and highly beneficial insecticides exert their actions.

Acknowledgements

The authors are indebted to Mrs Y. Matsumura and Mrs V. Rule for secretarial assistance in the preparation of this review. We are also indebted to many colleagues for generously supplying reprints of their research papers. Our particular thanks are due to Dr Norman Janes of Rothamsted Experimental Station, Harpenden, U.K. for advice on chemical nomenclature. Any errors or inconsistencies that may remain are entirely the responsibility of the authors.

References

Abalis, I. M., Eldefrawi, M. E. and Eldefrawi, A. T. (1983). Interaction of pyrethroids with GABA/benzodiazepine receptors in insect muscles and rat brain. *Fed. Proc.* **42**, 753.
Abalis, I. M., Eldefrawi, M. E. and Eldefrawi, A. T. (1983b). Biochemical identifica-

tion of putative GABA/benzodiazepine receptors in house fly thorax muscles. *Pestic. Biochem. Physiol.* **20**, 39–48.

Abalis, I. M., Eldefrawi, M. E. and Eldefrawi, A. T. (1986). Effects of insecticides on GABA-induced chloride influx into rat brain microsacs. *J. Toxicol. Env. Health* **18**, 13–23.

Abassy, M. A., Eldefrawi, M. E. and Eldefrawi, A. T. (1982). Allethrin interactions with the nicotinic acetylcholine receptor channel. *Life Sci.* **31**, 1547–1552.

Abassy, M. A., Eldefrawi, M. E. and Eldefrawi, A. T. (1983a). Pyrethroid action on the nicotinic acetylcholine receptor/channel. *Pestic. Biochem. Physiol.* **19**, 299–308.

Abassy, M. A., Eldefrawi, M. E. and Eldefrawi, A. T. (1983b). Influence of the alcohol moiety of pyrethroids on their interactions with the nicotinic acetylcholine receptor. *J. Toxicol. Env. Health.* **12**, 575–590.

Abdel-Aal, Y. A. I. and Soderlund, D. M. (1980). Pyrethroid hydrolysing esterases in southern armyworm *Spodoptera eridania* larvae: tissue distribution, kinetic properties and selective inhibition. *Pestic. Biochem. Physiol.* **14**, 282–289.

Adams, M. E. and Miller, T. A. (1979a). Site of action of pyrethroids: repetitive "backfiring" in flight motor units of house fly. *Pestic. Biochem. Physiol.* **11**, 218–231.

Adams, M. E. and Miller, T. A. (1979b). Neurophysiological correlates of pyrethroid and DDT-type insecticide poisoning in the house fly, *Musca domestica* L. *In* "Insect Neurobiology and Pesticide Action", pp. 431–439. Society for Chemical Industry, London.

Adams, M. E. and Miller, T. A. (1980). Neural and behavioural correlates of pyrethroid and DDT type poisoning in the house fly *Musca domestica. Pestic. Biochem. Physiol.* **13**, 137–147.

Ahn, Y-J., Shono, T. and Fukami, J. (1986a). Linkage group analysis of nerve insensitivity in a pyrethroid resistant strain of housefly. *Pestic. Biochem. Physiol.* **26**, 231–237.

Ahn, Y-J., Shono, T. and Fukami, J. (1986b). Inheritance of pyrethroid resistance in a housefly strain from Denmark. *J. Pestic. Sci.* **11**, 591–596.

Aldridge, W. N., Clothier, B., Forshaw, P., Johnson, M. K., Parker, V. H., Price, R. J., Skilleter, D. N., Verschoyle, R. D. and Stevens, C. (1978). The effect of DDT and the pyrethroids cismethrin and decamethrin on the acetylcholine and cyclic nucleotide content of rat brain. *Biochem. Pharmacol.* **27**, 1703–1706.

Ashcroft, F. M. and Stanfield, P. R. (1982). Calcium and potassium currents in muscle fibres of an insect (*Carausius morosus*). *J. Physiol. (London)* **323**, 93–115.

Bakry, N., Atteia, A., Aly, H. and Abdel-Ghaney, M. (1984). Effect of synthetic pyrethroids on mitochondrial Mg^{2+}-ATPase of *Spodoptera littoralis. J. Egypt. Soc. Toxicol.* **1**, 87–100.

Benos, D. J., Saccomani, G., Brenner, B. M. and Sariban-Sohraby, S. (1986). Purification and characterization of the amiloride-sensitive sodium channel from A6 cultured cells and bovine renal papilla. *Proc. Natl. Acad. Sci. U.S.A.* **83**, 8525–8529.

Berlin, J. R., Akera, T., Brody, T. M. and Matsumura, F. (1984). The inotropic effects of a synthetic pyrethroid decamethrin on isolated guinea pig atrial muscle. *Eur. J. Pharmacol.* **98**, 313–322.

Blagburn, J. M. and Sattelle, D. B. (1987a). Presynaptic depolarization mediates presynaptic inhibition at a synapse between an identified mechanosensory neurone and giant interneurone 3 in the first instar cockroach *Periplaneta americana. J. exp. Biol.* **127**, 135–157.

Blagburn, J. M. and Sattelle, D. B. (1987b). Calcium conductance in an identified cholinergic synaptic terminal in the central nervous system of the cockroach *Periplaneta americana*. *J. exp. Biol.* **129**, 347–364.

Blagburn, J. M. and Sattelle, D. B. (1987c). Nicotinic acetylcholine receptors on a cholinergic nerve terminal in the cockroach *Periplaneta americana*. *J. comp. Physiol.* **160** (in press).

Bloomquist, J. R., Adams, P. M. and Soderlund, D. M. (1986). Inhibition of γ-aminobutyric acid-stimulated chloride flux in mouse brain vesicles by polychlorocycloalkane and pyrethroid insecticides. *Neurotoxicology* **7**, 11–20.

Bloomquist, J. R. and Miller, T. A. (1985). Carbofuran triggers flight motor output in pyrethroid-blocked reflex pathways of the house fly. *Pestic. Biochem. Physiol.* **23**, 247–255.

Bloomquist, J. R. and Miller, T. A. (1986a). Sodium channel neurotoxins as probes of the knockdown resistance mechanism. *Neurotoxicology* **7**, 217–224.

Bloomquist, J. R. and Miller, T. A. (1986b). Neural correlates of flight activation and escape behaviour in houseflies recovering from pyrethroid poisoning. *Arch. Insect Biochem. Physiol.* **3**, 551–560.

Bloomquist, J. R. and Soderlund, D. M. (1985). Neurotoxic insecticides inhibit GABA-dependent chloride uptake by mouse brain vesicles. *Biochem. Biophys. Res. Commun.* **133**, 37–43.

Blum, M. S. and Kearns, C. W. (1956). Temperature and the action of pyrethrum in the American cockroach. *J. Econ. Entomol.* **49**, 862–865.

Bradbury, J. E., Forshaw, P. J., Gray, A. J. and Ray, D. E. (1983). The action of mephensin and other agents on the effects produced by two neurotoxic pyrethroids in the intact and spinal rat. *Neuropharmacology* **22**, 907–914.

Braestrup, C. and Squires, R. F. (1977). Specific benzodiazepine receptors in rat brain characterized by high affinity ^{3}H diazepam binding. *Proc. Natl. Acad. Sci.* **74**, 3805–3809.

Breer, H. and Heilgenberg, H. (1985). Neurochemistry of GABAergic activities in the central nervous system of *Locusta migratoria*. *J. Comp. Physiol.* **A157**, 343–354.

Breer, H., Kleene, R. and Hinz, G. (1985). Molecular forms and subunit structure of the acetylcholine receptor in the central nervous system of insects. *J. Neurosci.* **5**, 3386–3392.

Breer, H. and Knipper, M. (1984). Characterization of acetylcholine release from insect synaptosomes. *Insect Biochem.* **14**, 337–344.

Briggs, G. G., Elliott, M., Farnham, A. W. and Janes, N. F. (1974). Structural aspects of the knockdown of pyrethroids. *Pestic. Sci.* **5**, 643–650.

Brodie, M. E. (1983). Correlations between cerebellar cyclic GMP and motor effects induced by deltamethrin: independence of olivo-cerebellar tract. *Neurotoxicology* **4**, 1–12.

Brodie, M. E. and Aldridge, W. N. (1982). Elevated cerebellar cyclic GMP levels during the deltamethrin-induced motor syndrome. *Neurobehav. Toxicol. and Teratol.* **4**, 109–113.

Burt, P. E., Elliott, M., Farnham, A. W., Janes, N. F., Needhom, P. H. and Pulman, D. A. (1974). Geometrical and optical isomers of 2,2-dimethyl-3 (2,2-dichlorovinyl) cyclopropane carboxylic acid and insecticidal esters with 5-benzyl-3-furyl methyl and 3-phenoxygenzyl alcohols. *Pestic. Sci.* **5**, 791–799.

Burt, P. E. and Goodchild, R. E. (1974). Knockdown by pyrethroids: its role in the intoxication process. *Pestic. Sci.* **5**, 625–633.

Cahn, R. S., Ingold, C. and Prelog, V. (1956). The specification of asymmetric configuration in organic chemistry. *Experientia* **12**, 81–124.

Campbell, W. C., Fischer, M. H., Stapley, E. O., Albers-Schonberg, G. and Jacob, T. A. (1983). Ivermectin: a potent new antiparasitic agent. *Science* **221**, 823–828.

Casida, J. E. (1983). Novel aspects of metabolism of pyrethroids. *In* "Pesticide Chemistry, Human Welfare and the Environment. Proc. 5th Int. Congr. Pestic. Chem., Kyoto Japan" (Eds J. Miyamoto and P. C. Kearney). Vol. 2, pp. 187–192. Pergamon Press, Oxford and New York.

Casida, J. E., Gammon, D. W., Glickman, A. H. and Lawrence, L. J. (1983). Mechanism of selective action of pyrethroid insecticides. *Ann. Rev. Pharmacol. Toxicol.* **23**, 413–438.

Catterall, W. A. (1986). Molecular properties of voltage-sensitive sodium channels. *Ann. Rev. Biochem.* **55**, 953–985.

Chalmers, A. E., Miller, T. A. and Olsen, R. W. (1987). Deltamethrin: a neurophysiological study of the sites of action. *Pestic. Biochem. Physiol.* **27**, 36–41.

Chalmers, A. E. and Osborne, M. P. (1986a). The crayfish stretch receptor organ: a useful model system for investigating the effects of neuroactive substances. I. The effect of DDT and pyrethroids. *Pestic. Biochem. Physiol.* **26**, 128–138.

Chalmers, A. E. and Osborne, M. P. (1986b). The crayfish stretch receptor organ: a useful model system for investigating the effects of neuroactive substances. II. A pharmacological investigation of pyrethroid mode of action. *Pestic. Biochem. Physiol.* **26**, 139–149.

Chang, C. P. and Plapp, F. W. Jr (1983a). DDT and pyrethroids: receptor binding and mode of action in the house fly *Musca domestica*. *Pestic. Biochem. Physiol.* **20**, 76–85.

Chang, C. P. and Plapp, F. W. Jr (1983b). DDT and pyrethroids: receptor binding in relation to knockdown resistance (*kdr*) in the housefly *Musca domestica*. *Pestic. Biochem. Physiol.* **20**, 86–91.

Chialiang, C. and Devonshire, A. L. (1982). Changes in membrane phospholipids identified by Arrhenius plots of acetylcholinesterase and associated with pyrethroid resistance (*kdr*) in the house fly. *Pestic. Biochem. Physiol.* **20**, 86–91.

Chinn, K. and Narahashi, T. (1986). Stabilization of sodium channel states by deltamethrin in mouse neuroblastoma cells. *J. Physiol. (London)* **380**, 191–208.

Chinn, K. and Narahashi, T. (1987). Effect of temperature on sodium channel states stabilized by deltamethrin. *Biophys. J.* **51**, 8a.

Clark, J. M. and Matsumura, F. (1982). Two different types of inhibitory effects of pyrethroids on nerve Ca- and Ca + Mg-ATPase activity in the squid, *Loligo pealei*. *Pestic. Biochem. Physiol.* **18**, 180–190.

Claudio, T., Ballivet, M., Patrick, J. and Heinemann, S. (1983). Nucleotide and deduced amino acid sequences of *Torpedo californica* acetylcholine receptor γ-subunit. *Proc. Natl. Acad. Sci. U.S.A.* **80**, 1111–1115.

Clements, A. N. and May, T. E. (1977). The actions of pyrethroids upon the peripheral nervous system and associated organs in the locust. *Pestic. Sci.* **8**, 661–680.

Cohen, E. and Casida, J. E. (1986). Effects of insecticides and GABAergic agents on housefly [^{35}S]t-butylbicyclophosphorothionate binding site. *Pestic. Biochem. Physiol.* **25**, 63–72.

Cole, L. M., Lawrence, L. J. and Casida, J. E. (1984) Similar properties of [^{35}S]t-butylbicyclophosphorothionate receptor and coupled components of the GABA receptor–ionophore complex in the brains of human, cow, rat, chicken and fish. *Life Sci.* **35**, 1755–1762.

Corbett, J. R., Wright, K. and Baillie, A. C. (1984). "The Biochemical mode of action of pesticides", pp. 1–382. Academic Press, London.

Conti-Tronconi, B. M. and Raftery, M. A. (1982). The nicotinic cholinergic receptor: correlation of molecular structure with functional properties. *Ann. Rev. Biochem.* **51**, 491–530.

Cremer, J. E. (1983). The influence in mammals of the pyrethroid insecticides. *In* "Developments in the Science and Practice of Toxicology" (Eds A. W. Hayes, R. C. Schnell and T. S. Miya), pp. 61–72. Elsevier Science B.V., Amsterdam.

Cremer, J. E., Cunningham, V. J., Ray, D. E. and Gurcharan, S. S. (1980). Regional changes in brain glucose utilization in rats given a pyrethroid insecticide. *Brain Res.* **194**, 278–282.

Cremer, J. E., Ray, D. E., Gurcharan, S. S. and Cunningham, V. J. (1981). A study of the kinetic behaviour of glucose based on simultaneous estimates of influx and phosphorylation in brain regions of rats in different physiological states. *Brain Res.* **221**, 331–342.

Cremer, J. E. and Seville, M. P. (1982). Comparative effects of two pyrethroids, deltamethrin and cismethrin on plasma catecholamines and on blood glucose and lactate. *Toxicol. Appl. Pharmacol.* **66**, 124–133.

Crombie, L. (1980). Chemistry and biosynthesis of natural pyrethrins. *Pestic. Sci.* **11**, 102–118.

David, J. A. and Sattelle, D. B. (1984). Actions of cholinergic pharmacological agents on the cell body membrane of the fast coxal depressor motoneurone of the cockroach (*Periplaneta americana*). *J. Exp. Biol.* **108**, 119–136.

Davies, J. H. (1985). The pyrethroids: an historical introduction. *In* "Pyrethroid Insecticides" (Ed. J. P. Leahey), pp. 1–41. Taylor and Francis, London and Philadelphia.

Desaiah, D. and Cat Komp, L. K. (1972). Sensitivity of the cockroach *Periplaneta americana* ATPases to pyrethrum. *Ann. Meet. Entomol. Soc. Amer.* p. 30.

De Vries, D. H. and Georghiou, G. P. (1981a). Decreased nerve sensitivity and decreased cuticular penetration as mechanisms of resistance to pyrethroids in 1R trans permethrin selected strain of the house fly *Musca domestica. Pestic. Biochem. Physiol.* **15**, 234–241.

De Vries, D. H. and Georghiou, G. P. (1981b). Absence of enhanced detoxication of permethrin in pyrethroid resistant house flies *Musca domestica. Pestic. Biochem. Physiol.* **15**, 242–252.

Doherty, J. D., Nishimura, K., Kurihara, N. and Fujita, T. (1986). Quantitative structure-activity studies of substituted benzyl chrysanthemates. 9. Calcium uptake inhibition in crayfish nerve cord and lobster axon homogenates *in vitro* by synthetic pyrethroids. *Pestic. Biochem. Physiol.* **25**, 295–305.

Dolphin, A. C. (1984). GABA$_B$ receptors: has adenylate cyclase inhibition any functional relevance? *Trends in Neuroscience* **7**, 363–364.

Duclohier, H. and Georgescauld, D. (1979). The effects of the insecticide decamethrin on action potential and voltage-clamp currents of *Myxicola* giant axon. *Comp. Biochem. Physiol.* **62**, 217–224.

Dunlap, K. and Fischbach, G. D. (1981). Neurotransmitters decrease the calcium conductance activated by depolarization of embryonic chick sensory neurones. *J. Physiol.* **317**, 519–535.

Eldefrawi, A. T. (1976). The acetylcholine receptor and its interaction with insecticides. *In* "Insecticide Biochemistry and Physiology" (Ed. C. F. Wilkinson), pp. 297–326. Plenum Press, New York.

Eldefrawi, M. E., Abassy, M. A. and Eldefrawi, A. T. (1983). Effects of environmental toxicants on nicotinic acetylcholine receptors: action of pyrethroids. *In* "Cellular and Molecular Neurotoxicology" (Ed. T. Narahashi), pp. 179–189. Raven Press, New York.

Eldefrawi, A. T., Mansour, N. A. and Eldefrawi, M. E. (1982). Insecticides affecting acetylcholine receptor interactions. *Pharmac. Ther.* **16**, 45–65.

Eldefrawi, M. E., Sherby, S. M., Abalis, I. M. and Eldefrawi, A. T. (1985). Interactions of pyrethroid and cyclodiene insecticides with nicotinic acetylcholine and GABA receptors. *Neurotoxicology* **6**, 47–62.

Elliott, M. (1971). The relationship between the structure and activity of pyrethroids. *Bull. Wld. Hlth. Org.* **44**, 315–324.

Elliott, M. (1977). Synthetic pyrethroids. *In* "Synthetic Pyrethroids" (Ed. M. Elliott). ACS Symposium Series No. 42, pp. 1–28. American Chemical Society Washington, D.C.

Elliott, M. (1985). Lipophilic insect control agents. *In* "Recent Advances in the Chemistry of Insect Control" (Ed. N. F. Janes), pp. 73–102. The Royal Society of Chemistry, London.

Elliott, M. and Janes, N. F. (1973). Chemistry of the natural pyrethrins. *In* "Pyrethrums" (Ed. J. E. Casida), pp. 56–100. Academic Press, New York and London.

Elliott, M. and Janes, N. F. (1977). Synthetic Pyrethroids. *In* "Synthetic Pyrethroids" (Ed. M. Elliott), pp. 29–48, ACS Symposium Series No. 42. American Chemical Society, Washington, D.C.

Elliott, M. and Janes, N. F. (1978a). Synthetic pyrethroids—a new class of insecticide. *Chem. Soc. Rev.* **7**, 473–505.

Elliott, M. and Janes, N. F. (1978b). Recent structure–activity correlations in synthetic pyrethroids. *In* "Advances in Pesticide Science. Pt 2" (Ed. H. Geissbuhler), pp. 166–173. Pergamon Press, Oxford and New York.

Elliott, M., Janes, N. F. and Pulman, D. A. (1974). The pyrethrins and related compounds. Part XVIII. Insecticidal 2,2-dimethylcyclopropanecarboxylates with new unsaturated 3-substituents. *J. Chem. Soc., Perkin Trans.* **1**, 2470–2474.

Elliott, M., Janes, N. F. and Potter, C. (1978). The future of pyrethroids in insect control. *Ann. Rev. Ent.* **23**, 443–469.

Enna, S. J. and Gallagher, J. P. (1983). Biochemical and electrophysiological characteristics of mammalian GABA receptors. *Int. Rev. Neurobiol.* **24**, 181–212.

Evans, M. H. (1976). End-plate potentials in frog muscle exposed to a synthetic pyrethroid. *Pestic. Biochem. Physiol.* **6**, 547–550.

Farnham, A. W. (1977). Genetics of resistance of houseflies (*Musca domestica* L.) to pyrethroids. I. Knock-down resistance. *Pestic. Sci.* **8**, 631–636.

Fischer, J. B. and Olsen, R. W. (1986). Biochemical aspects of GABA/benzodiazepine receptor function. *In* "Receptor Biochemistry and Methodology. Vol. 5" (Eds. J. C. Venter and L. C. Harrison), pp. 241–259.

Ford, M. (1979). Quantatitive structure–activity relationships of pyrethroid insecticides. *Pestic. Sci.* **10**, 39–49.

Fujitani, J. (1909). Chemistry and pharmacology of insect powder. *Arch. Exp. Path. Pharmak.* **61**, 47–75.

Gallagher, J. P. and Shinnick-Gallagher, P. (1983). Electrophysiological characteristics of GABA-receptor complexes. *In* "The GABA Receptors" (Ed. S. J. Enna), pp. 25–61. Humana Press.

Gallagher, J. P., Nakamura, J. and Shinnick-Gallagher, P. (1983). The effects of temperature, pH and Cl-pump inhibitors on GABA responses recorded from cat dorsal root ganglia. *Brain Res.* **267**, 249–259.

Gammon, D. W. (1977). Nervous effects of toxins on an intact insect: a method. *Pestic. Biochem. Physiol.* **7**, 1–7.

Gammon, D. W. (1978). Neural effects of allethrin on the free walking cockroach *Periplaneta americana*: an investigation using defined doses at 15 and 32°C. *Pestic. Sci.* **9**, 79–91.

Gammon, D. W. (1979). An analysis of the temperature-dependence of the toxicity of allethrin to the cockroach. *In* "Neurotoxicology of Insecticides and Pheromones" (Ed. T. Narahashi), pp. 97–117. Plenum Press, New York.

Gammon, D. W. (1980). Pyrethroid resistance in a strain of *Spodoptera littoralis* is correlated with decreased sensitivity of the central nervous system *in vitro*. *Pestic. Biochem. Physiol.* **13**, 52–62.

Gammon, D. W. (1985). Correlations between *in vitro* and *in vivo* mechanisms of pyrethroid insecticide action. *Fund. Appl. Toxicol.* **5**, 9–23.

Gammon, D. W., Brown, M. A. and Casida, J. E. (1981). Two classes of pyrethroid action in the cockroach *Periplaneta americana*. *Pestic. Biochem. Physiol.* **15**, 181–191.

Gammon, D. W. and Casida, J. E. (1983). Pyrethroids of the most potent class antagonize GABA action at the crayfish neuromuscular junction. *Neurosci. Lett.* **40**, 163–168.

Gammon, D. W., Ruzo, L. O. and Casida, J. E. (1983). A new pyrethroid insecticide with remarkable potency on nerve axons. *Neurotoxicology* **4**, 165–170.

Gammon, D. W. and Sander, G. (1985). Two mechanisms of pyrethroid action: electrophysiological and pharmacological evidence. *Neurotoxicology* **6**, 63–86.

Ganetzky, B. and Wu, C-F. (1986). Neurogenetics of membrane excitability in *Drosophila*. *Ann. Rev. Genet.* **20**, 13–44.

Ghiasuddin, S. M. and Soderlund, D. M. (1984). Mouse brain synaptosomal sodium channels: activation by aconitine, batrachotoxin and veratridine, and inhibition by tetrodotoxin. *Comp. Biochem. Physiol.* C **77**, 267–271.

Ghiasuddin, S. M. and Soderlund, D. M. (1985). Pyrethroid insecticides: potent, stereospecific enhancers of mouse brain sodium channel activation. *Pestic. Biochem. Physiol.* **24**, 200–206.

Glickman, A. H. and Casida, J. E. (1982). Species and structural variations affecting pyrethroid neurotoxicity. *Neurobehav. Toxicol. Teratol.* **4**, 793–799.

Goodman, C. S. and Spitzer, N. C. (1981). The mature electrophysiological properties of identified neurones in grasshopper embryos. *J. Physiol. (London)* **313**, 369–384.

Guillet, J. C., Roche, M. and Pichon, Y. (1986). Effects of topical and bath applications of the insecticide deltamethrin on electrical activity in a leg mechanoreceptor of the cockroach, *Periplaneta americana*. *Pestic. Biochem. Physiol.* **26**, 183–192.

Hall, L. M., Kasbekar, D. P., Gil, D. W., Keen, J. K., Urquhart, D., Nelson, J. C. and Jackson, F. R. (1986). Molecular and genetic analysis of voltage-sensitive sodium channels from *Drosophila melanogaster*. *In* "Abstracts of papers presented at the 1986 meeting on Molecular Neurobiology of Drosophila", p. 27. Cold Spring Harbor Laboratory.

Hamill, O. P., Marty, A., Neher, E., Sakmann, B. and Sigworth, F. J. (1981). Improved patch-clamp techniques for high resolution current recording from cells and cell free membrane patches. *Pflügers Arch.* **391**, 85–100.

Hanke, W. and Breer, H. (1986). Channel properties of an insect neuronal acetylcholine receptor protein reconstituted in planar lipid bilayers. *Nature* **321**, 171–174.

Harris, R. A. and Allan, A. M. (1985). Functional coupling of γ-aminobutyric acid receptors to chloride channels in brain membranes. *Science* **228**, 1108–1110.

Heidmann, T. and Changeux, J. P. (1978). Structural and functional properties of the acetylcholine receptor protein in its purified and membrane-bound states. *Ann. Rev. Biochem.* **47**, 317–357.

Hendy, C. H. and Djamgoz, M. B. A. (1985). Effects of deltamethrin on ventral nerve cord activity in the cockroach. *Pestic. Sci.* **16**, 520–529.

Hille, B. (1984). "Ionic Channels of Excitable Membranes", pp. 1–426. Sinauer Sunderland, Massachusetts, U.S.A.

Hodgkin, A. L. and Huxley, A. F. (1952). A quantitative description of membrane current and its application to conduction and excitation in nerve. *J. Physiol. (London)* **117**, 500–544.

Hodgson, E. and Tate, L. G. (1976). Cytochrome P450 interactions. *In* "Insecticide Biochemistry and Physiology" (Ed. C. F. Wilkinson), pp. 115–148. Heyden, London.

Holloway, S. F., Salgado, V. L., Wu, C. H. and Narahashi, T. (1984). Maintained opening of single Na^+ channels by fenvalerate. *Neurosci. Abstracts* **10**, 864.

Horn, R. and Patlak, J. B. (1980). Single channel currents from excised patches of muscle membrane. *Proc. Natl. Acad. Sci. U.S.A.* **77**, 6930–6934.

Horn, R., Vandenberg, C. A. and Lange, K. (1984). Statistical analysis of single sodium channels. Effects of *N*-bromoacetamide. *Biophys. J.* **45**, 323–335.

Huang, L.-Y.-M., Moran, N. and Ehrenstein, G. (1984). Gating kinetics of batrachotoxin-modified sodium channels in neuroblastoma cells determined from single-channel measurements. *Biophys. J.* **45**, 313–322.

Hue, B. (1987). Actions of deltamethrin on synaptic transmission in the cockroach *Periplaneta americana*. *Comp. Biochem. Physiol.* (in press).

Hutzel, J. M. (1942). Action of pyrethrum upon the German cockroach. *J. Econ. Entomol.* **35**, 933–937.

Irving, S. N. (1984). *In vitro* activity of pyrethroids. *British Crop Protection Conference—Pests and Diseases* **9A-3**, 859–864.

Ishaaya, I. and Casida, J. E. (1980). Properties and toxicological significance of esterases hydrolysing permethrin and cypermethrin in *Trichoplusia-ni* larval gut and integument. *Pestic. Biochem. Physiol.* **14**, 178–184.

Ishaaya, I. and Casida, J. E. (1981). Pyrethroid esterases may contribute to natural pyrethroid tolerance of larvae of the common green lacewing *Chrysopa carnea*. *Environ. Entomol.* **10**, 681–684.

Jacques, Y., Romey, G., Cavey, M. T., Kartalovski, I-B. and Lazdunski, M. (1980). Interaction of pyrethroids with the Na^+ channel in mammalian cells in culture. *Biochim. Biophys. Acta* **600**, 882–897.

Janes, N. F. (1985). (Ed.) "Recent Advances in the Chemistry of Insect Control", pp. 1–322. Royal Society of Chemistry, London.

Jones, O. T. and Lee, A. G. (1986). Effects of pyrethroids on the activity of a purified calcium-magnesium ATPase. *Pestic. Biochem. Physiol.* **25**, 420–430.

Kubo, T., Fukuda, K., Mikami, A., Maeda, A., Takahashi, H., Mishina, M., Haga, T., Haga, K., Ichiyama, A., Kanagawa, K., Kokjima, M., Matsuo, H., Hirose, T. and Numa, S. (1986). Cloning, sequencing and expression of complementary DNA encoding the muscarinic acetylcholine receptor. *Nature (London)* **323**, 411–416.

Kuffler, S. W. and Edwards, C. (1958). Mechanisms of gamma aminobutyric acid (GABA) action and its relation to synaptic inhibition. *J. Neurophysiol.* **21**, 589–610.

Laufer, J., Roche, M., Pelhate, M., Elliot, M., Janes, N. F. and Sattelle, D. B. (1984).

Pyrethroid insecticides: actions of deltamethrin and related compounds on insect axonal sodium channels. *J. Insect Physiol.* **30**, 341–349.

Laufer, J., Pelhate, M. and Sattelle, D. B. (1985). Actions of pyrethroid insecticides on insect axonal sodium channels. *Pestic. Sci.* **16**, 651–661.

Lawrence, L. J. and Casida, J. E. (1982). Pyrethroid toxicology: mouse intracerebral structure-toxicity relationships. *Pestic. Biochem. Physiol.* **18**, 9–14.

Lawrence, L. J. and Casida, J. E. (1983). Stereospecific action of pyrethroid insecticides on the γ-aminobutyric acid receptor–ionophore complex. *Science* **221**, 1399–1401.

Lawrence, L. J., Gee, K. W. and Yamamura, H. I. (1985). Interactions of pyrethroid insecticides with chloride ionophore-associated binding sites. *Neurotoxicology* **6**, 87–98.

Leahey, J. P. (1985). (Ed.). "The Pyrethroid Insecticides", pp. 1–440. Taylor and Francis, London and Philadelphia.

Leake, L. D. (1982). Do pyrethroids activate neurotransmitter receptors? *Comp. Biochem. Physiol.* **72C**, 317–323.

Leake, L. D., Buckley, D. S., Ford, M. G. and Salt, D. W. (1985). Comparative effects of pyrethroids on neurones of target and non-target organisms. *Neurotoxicology* **6**, 99–116.

Leeb-Lundberg, F. and Olsen, R. W. (1980). Picrotoxin binding as a probe of the GABA postsynaptic membrane receptor-ionophore complex. *In* "Psychopharmacology and Biochemistry of Neurotransmitter Receptors" (Ed. H. I. Yamamura, R. W. Olsen and E. Usdin), pp. 593–606. Elsevier/North Holland, New York.

Lhoste, J. (1964). Les pyrethrins. *Phytoma, defense des cultures* **161**, 21–25.

Lummis, S. C. R., Chen-Chow, S., Holan, G. and Johnston, G. A. R. (1987a). γ-aminobutyric acid receptor ionophore complexes: differential effects of a rat brain membrane preparation. *J. Neurochem.* **48**, 689–694.

Lummis, S. C. R., Holan, G., Sattelle, D. B. and Johnston, G. A. R. (1987b). Insecticide effects on GABA-stimulated chloride flux in insect nervous tissue. *In* "Proceedings Xth Int. Congr. Pharmacol". IUPHAR, Sydney, Australia.

Lummis, S. C. R. and Sattelle, D. B. (1985a). Insect central nervous system γ-aminobutyric acid receptors. *Neurosci. Letts* **60**, 13–18.

Lummis, S. C. R. and Sattelle, D. B. (1985b). Binding of N-[propionyl-^{3}H] propionylated α-bungarotoxin and L-[benzilic-4,4'-^{3}H] quinuclidinyl benzilate to CNS extracts of the cockroach *Periplaneta americana*. *Comp. Biochem. Physiol.* **80C**, 75–83.

Lummis, S. C. R. and Sattelle, D. B. (1986). Binding sites for [^{3}H]GABA, [^{3}H]flunitrazepan and [^{35}S]TBPS in insect CNS. *Neurochem. Int.* **9**, 287–293.

Lund, A. E. and Narahashi, T. (1979). The effect of the insecticide tetramethrin on the sodium channel of crayfish giant axons. *Soc. Neurosci. Abstr.* **5**, 293.

Lund, A. E. and Narahashi, T. (1981a). Modification of sodium channel kinetics by the insecticide tetramethrin in crayfish giant axons. *Neurotoxicology* **2**, 213–229.

Lund, A. E. and Narahashi, T. (1981b). Kinetics of sodium channel modification by the insecticide tetramethrin in squid axon membranes. *J. Pharmacol. exp. Ther.* **219**, 464–473.

Lund, A. E. and Narahashi, T. (1982). Dose-dependent interaction of the pyrethroid isomers with sodium channels of squid axon membranes. *Neurotoxicology* **3**, 11–24.

Lund, A. E. and Narahashi, T. (1983). Kinetics of sodium channel modification as the basis for the variation in the nerve membrane effects of pyrethroids and DDT analogs. *Pestic. Biochem. Physiol.* **20**, 203–216.

Lunt, G. G., Robinson, T. N., Knowles, W. P. and Olsen, R. W. (1985). The identification of GABA receptor binding sites in insect ganglia. *Neurochem. Int.* **7,** 751–754.
McCleskey, E. W., Fox, A. P., Feldman, D. and Tsien, R. W. (1986). Different types of calcium channels. *J. exp. Biol.* **124,** 177–190.
Milani, R. (1954). Comportamento mendeliano della resistenza alla azione abbattente del DDT: correlazione tra abbattimento e mortalita in *Musca domestica* L. *Riv. Parassit.* **15,** 513–542.
Miller, R. J. (1987). Multiple channels and neuronal function. *Science* **235,** 46–52.
Miller, T. A., Kennedy, J. M. and Collins, C. (1979). Central nervous system insensitivity to pyrethroids in the resistant *kdr* strain of houseflies *Musca domestica*. *Pestic. Biochem. Physiol.* **12,** 224–230.
Miller, T. A. and Salgado, V. L. (1985). The mode of action of pyrethroids on insects. *In* "The Pyrethroid Insecticides" (Ed. J. P. Leahey), pp. 43–97. Taylor and Francis, London and Philadelphia.
Nagy, K., Kiss, T. and Hof, D. (1983). Single Na channels in mouse neuroblastoma cell membranes. Indications for two open states. *Pflügers Arch.* **399,** 302–308.
Narahashi, T. (1971). Effects of insecticides on exictable tissues. *Adv. Insect. Physiol.* **8,** 1–93.
Narahashi, T. (1974). Chemicals as tools in the study of excitable membranes. *Physiol. Rev.* **54,** 813–889.
Narahashi, T. (1980). Site and types of action of pyrethroids on nerve membrane. *In* "Pyrethroid Insecticides: Chemistry and Action" (Ed. J. Mathieu), pp. 15–17. Roussel-Uclaf, Romainville.
Narahashi, T. (1982). Cellular and molecular mechanisms of actions of insecticides: neurophysiological approach. *Neurobehav. Toxicol. Teratol.* **4,** 753–758.
Narahashi, T. (1985). Nerve membrane ionic channels as the primary target of pyrethroids. *Neurotoxicology* **6,** 3–22.
Narahashi, T. and Anderson, N. C. (1967). Mechanism of excitation block by the insecticide allethrin applied externally and internally to squid giant axons. *Toxicol. Appl. Pharmacol.* **10,** 529–547.
Neher, E. and Sakmann, B. (1976). Single-channel currents recorded from membrane of denervated frog muscle fibres. *Nature (London)* **260,** 779–802.
Naumann, K. (1981). Chemie der synthetischen Pyrethroid-Insektizide. *In* "Chemie der Pflanzenschute und Schadlingsbekampfungsmittel. Vol. 7" (Ed. R. Wegler). Springer-Verlag, Berlin.
Newberg, N. R. and Nicoll, R. A. (1984a). Direct hyperpolarizing action of baclofen on hippocampal pyramidal cells. *Nature* **308,** 450–452.
Newberry, N. R. and Nicoll, R. A. (1984b). A bicuculline-resistant inhibitory postsynaptic potential in rat hippocampal pyramidal cells *in vitro*. *J. Physiol. (London)* **348,** 239–254.
Nishimura, K. and Narahashi, T. (1978). Structure activity relationships of pyrethroids based on direct action on nerve. *Pestic. Biochem. Physiol.* **8,** 53–64.
Nishimura, K., Okajima, N., Fujita, T. and Nakajima, M. (1982). Quantitative structure–activity studies of substituted benzyl chrysanthemates. 4. Physicochemical properties and the rate of progress of the knockdown symptom induced in house flies. *Pestic. Biochem. Physiol.* **18,** 341–350.
Noda, M., Ikeda, T., Kayano, T., Suzuki, H., Takeshima, H., Kurasaki, M., Takahashi, H. and Numa, S. (1986). Existence of distinct sodium channel messenger RNAs in rat brain. *Nature (London)* **320,** 188–192.
Noda, M., Takahashi, T., Tanabe, T., Toyosato, M., Furutani, Y., Hirose, T., Asai, M.,

Inayama, S., Miyata, T. and Numa, S. (1982). Primary structure of subunit precursors of *Torpedo californica* acetylcholine receptor deduced from cDNA sequence. *Nature (London)* **299**, 793–797.

Noda, M., Takahashi, H., Tanabe, M., Toyosato, M., Kiyotani, S., Furutani, Y., Hirose, T., Takashima, H., Inayama, S., Miyata, T. and Numa, S. (1983a). Structural homology of *Torpedo californica* acetylcholine receptor subunits. *Nature (London)* **302**, 528–532.

Noda, M., Takahashi, T., Tanabe, T., Toyosato, M., Kiyotani, S., Hirose, T., Asai, M., Takashima, H., Inayama, S., Miyata, T. and Numa, S. (1983b). Primary structure of β and δ-subunit precursors of *Torpedo californica* acetylcholine receptor deduced from cDNA sequences. *Nature (London)* **301**, 251–255.

Noda, M., Shimizu, S., Tanaka, T., Takai, T., Kayano, T., Ikeda, T., Takahashi, H., Nakayama, H., Kanaoka, Y., Minamino, N., Kangawa, K., Matsuo, H., Raftery, M. A., Hirose, T., Inayama, S., Hayashida, H., Miyata, T. and Numa, S. (1984). Primary structure of *Electrophorus electricus* sodium channel deduced from cDNA sequence. *Nature (London)* **312**, 121–127.

Nowycky, M. C., Fox, A. P. and Tsien, R. W. (1985). Three types of neuronal calcium channel with different calcium agonist sensitivity. *Nature (London)* **316**, 440–443.

Olsen, R. W. (1981). GABA-benzodiazepine–barbiturate receptor interactions. *J. Neurochem.* **37**, 1–13.

Orchard, I. (1976). Calcium dependent action potentials in a peripheral neurosecretory cell of the stick insect. *J. Comp. Physiol.* **112**, 95–102.

Orchard, I. (1980). Effects of pyrethroids on the electrical activity of neurosecretory cells from the brain of *Rhodnius prolixus*. *Pestic. Biochem. Physiol.* **13**, 220–226.

Orchard, I. and Osborne, M. P. (1977). The effects of cations upon the action potentials recorded from neurohaemal tissue of the stick insect. *J. Comp. Physiol.* **118**, 1–12.

Orchard, I. and Osborne, M. P. (1979). The action of insecticides on neurosecretory neurones in the stick insect *Carausius morosus*. *Pestic. Biochem. Physiol.* **10**, 197–202.

Osborne, M. P. and Smallcombe, A. (1983). Site of action of pyrethroid insecticides in neural membranes as revealed by the *kdr* resistance factor. *In* "Pesticide Chemistry Human Welfare and the Environment. Vol. 3" (Eds J. Miyamoto and P. C. Kearney), pp. 103–107. Pergamon Press, Oxford.

Otsuka, M., Iversen, L. L., Hall, Z. W. and Kravitz, E. A. (1966). Release of gamma-aminobutyric acid from inhibitory nerves of lobster. *Proc. Natl. Acad. Sci. U.S.A.* **56**, 1110–1115.

Pelhate, M. and Sattelle, D. B. (1982). Pharmacological properties of insect axons: a review. *J. Insect. Physiol.* **28**, 889–903.

Pinnock, R. D. and Sattelle, D. B. (1987). Dissociation and maintenance *in vitro* of neurones from adult cockroach (*Periplaneta americana*) and housefly (*Musca domestica*). *J. Neurosci. Methods* **20**, 195–202.

Pitman, R. M. (1979). Intracellular citrate and externally applied tetraethyl ammonium ions produce calcium-dependent action potentials in an insect motorneurone cell body. *J. Physiol. (London)* **291**, 327–337.

Plapp, F. W. Jr and Casida, J. E. (1969). Genetic control of house fly NADPH-dependent oxidases: relation to insecticide chemical metabolism and resistance. *J. Econ. Entomol.* **62**, 1174–1179.

Quandt, F. N. and Narahashi, T. (1982). Modification of single Na^+ channels by batrachotoxin. *Proc. Nat. Acad. Sci. U.S.A.* **79**, 6732–6736.

Ray, D. E. (1982). Changes in brain blood flow associated with deltamethrin-induced choreoathetosis in the rat. *Exp. Brain Res.* **45**, 269–276.

Ray, D. E. and Cremer, J. E. (1979). The action of decamethrin (a synthetic pyrethroid) on the rat. *Pestic. Biochem. Physiol.* **10**, 333–340.

Roberts, E., Krause, D. N., Wong, E. and Mori, E. (1976). Different efficacies of D and ʟ amino hydroxybutyric acids in GABA receptor and transport test systems. *J. Neurosci.* **1**, 132–140.

Robinson, T. N., Lunt, G. G., Battersby, M., Irving, S. and Olsen, R. W. (1986). Insect ganglia contain [^{3}H]flunitrazepam binding sites. *Biochem. Soc. Trans.* **13**, 716–717.

Roche, M., Frelin, C., Bruneau, P. and Meinard, C. (1985). Interaction of tralomethrin, tralocythrin, and related pyrethroids in Na$^+$ channels of insect and mammalian neuronal cells. *Pestic. Biochem. Physiol.* **24**, 306–316.

Romey, G., Chicheportiche, R. and Lazdunski, M. (1980). Transition temperatures of the electrical activity of ion channels in the nerve membrane. *Biochem. Biophys. Acta* **602**, 610–620.

Romey, G. and Lazdunski, M. (1982). Lipid-soluble toxins thought to be specific for Na$^+$ channels block Ca^{2+} channels in neuronal cells. *Nature (London)* **297**, 79–80.

Ruigt, G. S. F., Klis, J. F. L. and van den Bercken, J. (1986). Pronounced repetitive activity induced by the pyrethroid insecticide, fenfluthrin, in the slowly adapting stretch receptor neuron of the crayfish. *J. Comp. Physiol. A.* **159**, 43–53.

Ruigt, G. S. F. and van den Bercken, J. (1986). Action of pyrethroids on a nerve muscle preparation of the clawed frog *Xenopus laevis. Pestic. Biochem. Physiol.* **25**, 176–187.

Ruzo, L. O., Unai, T. and Casida, J. E. (1978). Decamethrin metabolism in rats. *J. Agric. Food Chem.* **26**, 918–925.

Salgado, V. L., Irving, S. N. and Miller, T. A. (1983a). Depolarization of motor nerve terminals by pyrethroids in susceptible and *kdr*-resistant house flies. *Pestic. Biochem. Physiol.* **20**, 100–114.

Salgado, V. L., Irving, S. N. and Miller, T. A. (1983b). The importance of nerve terminal depolarization in pyrethroid poisoning of insects. *Pestic. Biochem. Physiol.* **20**, 169–192.

Salgado, V. L. and Narahashi, T. (1983). Current–voltage relations of normal and fenvalerate-modified sodium channels in crayfish axons. *Biophys. J.* **41**, 51a.

Salgado, V. L. and Narahashi, T. (1988). Block of sodium channel gating current by the pyrethroid fenvalerate. *Biophys. J.* (in press).

Salkoff, L. B., Butler, A., Wei, A., Scavarda, N., Giffen, K., Ifune, C., Goodman, R. and Mandel, G. (1987). Genomic organization and deduced amino acid sequence of a putative sodium channel gene in *Drosophila. Science* **237**, 744–749.

Salkoff, L. B. and Wyman, R. J. (1983). Ion currents in *Drosphila* flight muscles. *J. Physiol. (London)* **337**, 687–709.

Sattelle, D. B. (1985). Acetylcholine receptors. *In* "Comprehensive Insect Physiology, Biochemistry and Pharmacology. Vol. 11 (Insect Pharmacology)" (Eds G. A. Kerkut and L. I. Gilbert), pp. 395–434. Pergamon Press, Oxford.

Sattelle, D. B. (1986). Acetylcholine receptors of insects: biochemical and physiological approaches. *In* "Neuropharmacology and Neurobiology" (Eds M. G. Ford, P. N. R. Usherwood, R. C. Reay and G. G. Lunt), pp. 445–497. Ellis-Horwood, Chichester, England.

Sattelle, D. B. (1988). Actions of deltamethrin on the responses to acetyl-choline and GABA of an identified insect motor neurone. *Neurosci. Letts.* (submitted for publication).

Sattelle, D. B. and Breer, H. (1985). Purification by affinity-chromatography of a nicotinic acetylcholine receptor from the CNS of the cockroach, *Periplaneta americana. Comp. Biochem. Physiol.* **82C,** 349–352.

Sattelle, D. B. and Callec, J. J. (1977). Actions of nereistoxin at an invertebrate central synapse. *Proc. Int. Congr. Physiol. Sci., Paris.* **Vol. XIII,** p. 662.

Sattelle, D. B., Harrow, I. D., David, J. A., Pelhate, M. and Callec, J. J. (1985). Nereistoxin: actions on a CNS acetylcholine receptor/ion channel in the cockroach *Periplaneta americana. J. exp. Biol.* **118,** 37–52.

Sattelle, D. B., Harrow, I. D., Hue, B., Pelhate, M., Gepner, J. I. and Hall, L. M. (1983). α-Bungarotoxin blocks excitatory synaptic transmission between cercal sensory neurons and giant interneurone 2 of the cockroach *Periplaneta americana. J. exp. Biol.* **107,** 473–489.

Sattelle, D. B., Pinnock, R. D., Wafford, K. A. and David, J. A. (1988). GABA receptors on the cell body membrane of an identified insect motor neurone. *Proc. Roy. Soc. London B.* **232** (in press).

Sattelle, D. B., Sun, Y. A. and Wu, C. F. (1986). Neuronal acetylcholine receptor: patch-clamp recording of single channel properties from dissociated insect neurones. *IRCS Med. Sci.* **14,** 65–66.

Sawicki, R. M. (1978). Unusual response of DDT-resistant houseflies to carbinol analogues of DDT. *Nature (London)* **875,** 443–444.

Sawicki, R. M. (1985). Resistance to pyrethroid insecticides in arthropods. *In* "Insecticides" (Ed. D. H. Hutson and D. R. Roberts), pp. 143–191. John Wiley, New York.

Schechter, M. S., Green, N. and La Forge, F. B. (1949). Constituents of pyrethrum flowers XXIII. Cineroline and the synthesis of related cyclopentlenones. *J. Amer. Chem. Soc.* **72,** 3541–3542.

Schouest, L. P., Salgado, V. L. and Miller, T. A. (1986). Synaptic vesicles are depleted from motor nerve-terminals of deltamethrin-treated house fly larvae, *Musca domestica. Pestic. Biochem. Physiol.* **25,** 381–386.

Scott, J. G. and Matsumura, E. (1983). Evidence for two types of toxic actions of pyrethroids on susceptible and DDT-resistant German cockroaches. *Pestic. Biochem. Physiol.* **19,** 141–150.

Seifert, J. and Casida, J. E. (1985a). Solubilization and detergent effects on interactions of some drugs and insecticides with the tert-butylbicyclophosphorothionate binding site within the gamma aminobutyric-acid receptor–ionophore complex. *J. Neurochem.* **44,** 110–116.

Seifert, J. and Casida, J. E. (1985b). Regulation of sulphur-35 tert-butylbicyclophosphorothionate binding sites in rat brain by gamma aminobutyric-acid pyrethroid and barbiturate. *Eur. J. Pharmacol.* **115,** 191–198.

Sherby, S. M., Eldefrawi, A. T., Deshpande, S. S., Albuquerque, E. X. and Eldefrawi, M. E. (1986). Effects of pyrethroids on nicotinic acetylcholine receptor binding and function. *Pestic. Biochem. Physiol.* **26,** 107–115.

Sibley, D. R. and Lefkowitz, R. J. (1985). Molecular mechanisms of receptor desensitization using the β-adrenergic receptor-coupled adenylate cyclase system as a model. *Nature (London)* **317,** 124–129.

Simmonds, M. A. (1983). Multiple GABA receptors and associated regulatory sites. *Trends in Neurosci.* **6,** 279–281.

Smith, P. R. (1980). The effect of cismethrin on the rat dorsal root potentials. *Eur. J. Pharmacol.* **66,** 125–128.

Soderlund, D. M. (1979). Pharmacokinetic behaviour of enantiomeric pyrethroid esters in the cockroach *Periplaneta americana*. *Pestic. Biochem. Physiol.* **12**, 38–48.

Soderlund, D. M. (1983). Pharmacokinetic approaches to the activity of pyrethroids in insects. *In* "Pesticide Chemistry, Human Welfare and the Environment. Vol. 3" (Eds J. Miyamoto and P. C. Kearney), pp. 69–74. Pergamon Press, Oxford and New York.

Soderlund, D. M. (1985). Pyrethroid–receptor interactions: stereospecific binding and effects on sodium channels in mouse brain preparations. *Neurotoxicology* **6**, 35–46.

Soderlund, D. M. and Casida, J. E. (1977). Substrate specificity of mouse-liver microsomal enzymes in pyrethroid metabolism. *In* Synthetic Pyrethroids" (Ed. M. Elliott). *ACS Symposium Series No. 42*, pp. 162–172. Am. Chem. Soc., Washington, D.C.

Soderlund, D. M., Ghiasuddin, S. M. and Helmuth, D. W. (1983). Receptor-like stereospecific binding of a pyrethroid insecticide to mouse brain membranes. *Life Sci.* **33**, 261–267.

Squires, R. F., Casida, J. E., Richardson, M. and Saederup, E. (1983). [^{35}S]t-Butylbicyclophosphorothionate binds with high affinity to brain specific sites coupled to γ-aminobutyric acid A and ion recognition sites. *Mol. Pharmacol.* **23**, 326–336.

Staatz, C. G., Bloom, A. S. and Lech, J. J. (1982). A pharmacological study of pyrethroid neurotoxicity in mice. *Pestic. Biochem. Physiol.* **17**, 287–292.

Staatz-Benson, C. G. and Hosko, M. J. (1986). Interaction of pyrethroids with mammalian spinal neurones. *Pestic. Biochem. Physiol.* **25**, 19–30.

Staudinger, H. and Ruzicka, L. (1924). Insektenstoffe Parts 7–10. *Helv. Chem. Acta* **7**, 390–458.

Supavilai, P. and Karobath, M. (1984). [^{35}S]t-butylcyclophosphorothionate binding sites are constituents of the γ-aminobutyric acid and benzodiazepine receptor complex. *J. Neurosci.* **4**, 1193–1200.

Suzuki, T. and Miyamoto, J. (1978). Purification and properties of pyrethroid carboxy esterase in rat liver microsomes. *Pestic. Biochem. Physiol.* **8**, 186–198.

Svestka, M. and Vankova, J. (1980). The effect of *Bacillus thuringiensis* combined with the synthetic pyrethroid Ambush on *Operophthera brumata*, *Tortrix viridana* and the Entomo fauna of an oak stand. *Anz Schaedlingsksd Pflanzenschutz Umweltschutz* **53**, 6–10.

Swenson, R. P. (1983). A slow component of gating current in crayfish giant axons resembles inactivation charge movement. *Biophys. J.* **41**, 245–249.

Takai, T., Noda, M., Mishina, M., Shimizu, S., Furutani, Y., Kayano, T., Ikeda, T., Kubo, T., Takashima, H., Takahashi, T., Kuno, M. and Numa, S. (1985). Cloning, sequencing and expression of cDNA for a novel subunit of acetylcholine receptor from calf muscle. *Nature (London)* **315**, 761–764.

Tessier, J. R. (1982). Stereochemical-specific synthesis of the most potent isomers. *In* "Deltamethrin, Chapter 2", pp. 38–44. Roussel-Uclaf.

Tessier, J. R. (1985). Stereochemical aspects of pyrethroid chemistry. *In* "Recent Advances in the Chemistry of Insect Control" (Ed. N. F. Janes), pp. 26–52. The Royal Society of Chemistry, London.

Thomas, M. V. (1984). Voltage-clamp analysis of calcium mediated potassium conductance in cockroach (*Periplaneta americana*) central neurones. *J. Physiol. (London)* **350**, 159–178.

Tsukamoto, M. and Casida, J. E. (1967). Metabolism of methylcarbamate insecticides by the NADPH$_2$-requiring enzyme system from houseflies. *Nature London* **213**, 49–51.

Usherwood, P. N. R. and Grundfest, H. (1965). Peripheral inhibition in skeletal muscle of insects. *J. Neurophysiol.* **28**, 497–518.

Vijverberg, H. P. M. and de Weille, J. R. (1985). The interaction of pyrethroids with voltage-dependent Na channels. *Neurotoxicology* **6**, 23–34.

Vijverberg, H. P. M. and van den Bercken, J. (1979). Frequency dependent effects of the pyrethroid insecticide decamethrin in frog *Xenopus laevis* myelinated nerve fibres. *Eur. J. Pharmacol.* **58**, 501–504.

Vijverberg, H. P. M., van der Zalm, J. M. and van den Bercken, J. (1982). Similar mode of action of pyrethroids and DDT on sodium channel gating in myelinated nerves. *Nature (London)* **295**, 601–603.

Vijverberg, H. P. M., van der Zalm, J. M., van Kleef, R. G. D. M. and van den Bercken, J. (1983). Temperature and structure-dependent interactions of pyrethroids with the sodium channels in frog node of Ranvier. *Biochim. Biophys. Acta* **728**, 73–82.

Vincent, J. P., Balerna, M., Barhanin, J., Fosset, M. and Lazdunski, M. (1980). Binding of sea anemone *Anemonia sulcata* toxin to receptor sites associated with gating system of sodium channel in synaptic nerve endings *in vitro*. *Proc. Natn. Acad. Sci. U.S.A.* **77**, 1646–1650.

de-Weille, J. R. and Vijverberg, H. P. M. and Narahashi, T. (1986). Sodium depletion in the periaxonal space of the squid *Loligo pealei* axon treated with pyrethroids. *Brain Res.* **386**, 169–174.

Wafford, K. A., Sattelle, D. B., Abalis, I., Eldefrawi, A. T. and Eldefrawi, M. E. (1987). γ-Aminobutyric acid-activated $^{36}Cl^{-}$ influx: a functional *in vitro* assay for CNS γ-aminobutyric acid receptors of insects. *J. Neurochem.* **48**, 177–180.

Wang, C. M., Narahashi, T. and Scuka, M. (1972). Mechanisms of negative temperature coefficient of nerve blocking action of allethrin. *J. Pharmacol. Exp. Ther.* **182**, 442–453.

Wing, K. D., Hammock, B. D. and Wustner, D. A. (1978). Development of an s-bioallethrin specific antibody. *J. Agric. Food Chem.* **26**, 1328–1333.

Wong, D. T., Threlheld, P. G., Bymaster, F. P. and Squires, R. F. (1984). Saturable binding of ^{35}S-t-butylbicyclophosphorothionate to the sites linked to the GABA receptor and the interaction with GABAergic agents. *Life Sci.* **34**, 853–860.

Wouters, W. and van den Bercken, J. (1978). Action of pyrethroids. *Gen. Pharmacol.* **9**, 387–398.

Wouters, W., van den Bercken, J. and van Gimeken, A. (1977). Presynaptic action of the pyrethroid insecticide allethrin in the frog motor endplate. *Eur. J. Pharmacol.* **43**, 163–171.

Yamamoto, D. (1985). The operation of the sodium channel in nerve and muscle. *Prog. Neurobiol.* **24**, 257–291.

Yamamoto, D. (1987). Sodium inward currents through calcium channels in mealworm muscle fibers. *Arch. Insect. Biochem. Physiol.* **5**, 227–231.

Yamamoto, D. and Fukami, J. (1977). Ionic requirements for non-synaptic electrogenesis in the muscle fibres of a lepidopterous insect. *J. exp. Biol.* **70**, 41–47.

Yamamoto, D., Fukami, J. and Washio, H. (1981). Voltage clamp studies on insect skeletal muscle. I. The inward current. *J. exp. Biol.* **92**, 1–12.

Yamamoto, D., Pinnock, R. D. and Sattelle, D. B. (1987). Switching between conductance states in a calcium-activated potassium channel of dissociated adult neurones of the cockroach (*Periplaneta americana*). *Proc. Natl. Acad. Sci. U.S.A.* submitted.

Yamamoto, D. and Narahashi, T. (1983). Two conductance states of open Na channels in neuroblastoma cells. *Fedn. Proc.* **42**, 599.

Yamamoto, D., Quandt, F. N. and Narahashi, T. (1983). Modification of single sodium channels by the insecticide tetramethrin. *Brain Res.* **274,** 344–349.

Yamamoto, D. and Washio, H. (1979). Permeation of sodium through calcium channels of an insect muscle membrane. *Can. J. Physiol. Pharmacol.* **57,** 220–222.

Yamamoto, D., Yeh, J. Z. and Narahashi, T. (1984). Voltage-dependent calcium block of normal and tetramethrin-modified single sodium channels. *Biophys. J.* **45,** 337–344.

Yamamoto, D., Yeh, J. Z. and Narahashi, T. (1985). Interactions of permeant cations with sodium channels of squid axon membranes. *Biophys. J.* **48,** 361–368.

Yamamoto, D., Yeh, J. Z. and Narahashi, T. (1986). Ion permeation and selectivity of squid axon sodium channels modified by tetramethrin. *Brain Res.* **372,** 193–197.

Index

Transverse nerve—*contd.*
  monoclonal antibodies, 93
    for staining, 113–114
  neuroendocrine cell, 88
    peripheral, differentiation, 103–110
    stereotyped development, 111–112
    targets, 112–113
  neurons in, 91, 92

Voltage clamp analysis of sodium channels, 164, 166–171
  and deltamethrin, 164, 166
  and fenvalerate, 169
  gating kinetics, 164, 166
  tail current amplitude, 168
  and tetramethrin, 166–168